# THE LOST BOYS

Gina Perry is an Australian psychologist and writer. She is author of *Behind the Shock Machine: the untold story of the notorious Milgram psychology experiments*, and was a co-producer of the ABC Radio National documentary 'Beyond the Shock Machine', which won the Silver World Medal for a history documentary in the 2009 New York Festivals radio awards. In 2013 she was runner up for the Bragg UNSW Prize for Science Writing, and her work has been anthologised in *Best Australian Science Writing* (2013 and 2015). Her feature articles, columns, and essays appear in publications including *The Age*, *The Australian*, and *Cosmos*. Gina has a PhD from the University of Melbourne, where she is an associate in the School of Culture and Communication. Learn more about Gina at www.gina-perry.com.

GINA PERRY

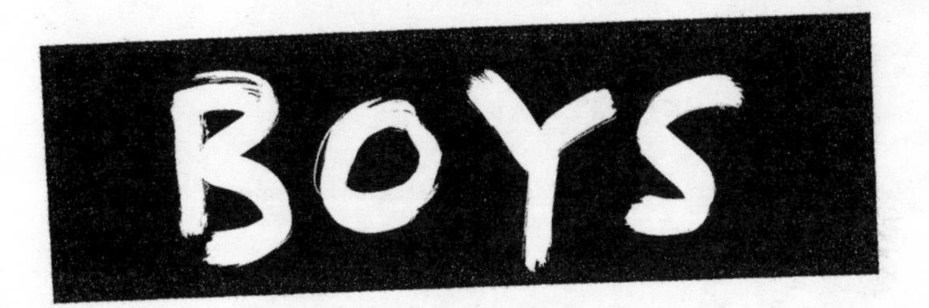


# THE LOST BOYS

## INSIDE MUZAFER SHERIF'S ROBBERS CAVE EXPERIMENT

SCRIBE
Melbourne • London

Scribe Publications
18–20 Edward St, Brunswick, Victoria 3056, Australia
2 John St, Clerkenwell, London, WC1N 2ES, United Kingdom

First published by Scribe 2018



Typeset in Adobe Garamond Pro 11.5/17pt by J&M Typesetting
Printed and bound in the UK by CPI Group (UK) Ltd, Croydon CR0 4YY

Scribe Publications is committed to the sustainable use of natural resources and the use of paper products made responsibly from those resources.

9781911344391 (UK edition)
9781925322354 (Australian edition)
9781925548303 (e-book)

CiP records for this title are available from the National Library of Australia and the British Library.

scribepublications.co.uk
scribepublications.com.au

# Contents

'No social science is more extravagantly
autobiographical than psychology.'

**Jill Lepore**

# Prologue

*Robbers Cave State Park, Oklahoma, 1954*

After the sun had gone down, the boys raced one another from the swimming hole to their cabins. They were still jubilant from their win, fizzing with excitement, eager to get back and pass around their prize again, the handsome silver knives fanned out on a stiff cardboard stand. Will shouted 'Told y'all!' triumphantly when he reached the cabin first. Panting and laughing, he threw open the cabin door — and stopped dead. Mattresses hung drunkenly from the bunks; pillows and clothes and comic books spilled across the floor. The knives, which they'd put on a makeshift table by the window, were gone. He let his breath out in a rush, then turned and started running, pushing past the group of dismayed boys who had crowded in behind him.

Outside, the long twilight was fading, and the stone huts dotted through the park were disappearing into the dark. He heard the others calling to him to wait up, but he didn't stop. He ran along the dusty track, feet pounding, and across the stream, his heart racing so hard he could hear his blood thrumming in his ears. Behind him the others had almost caught him up. The jumble of their voices quietened and the air was full of the sound of panting breaths. No need to stop and think, they just followed their instincts — an animal need to retrieve what was theirs. Will raced past the mess hall, where the sounds of a cowboy tune twanging on the radio and the clatter of dishes reminded him of home, of his parents' heads bowed as they said grace over supper. But he ran faster, thrusting those images behind him. His parents couldn't help him now. And prayers were no use here. But he wasn't afraid — he was as fierce as the soldiers and the Indians whose spirits whispered in the trees. When he first came here, he tried not to think about what animals were moving through the dark. Now he bared his teeth as he ran. Tonight he wouldn't be scared if a mountain lion stepped out of the shadows, or a bear climbed down from a tree. Behind him, the other boys rushed. They were a single panting pack, zigzagging in and out of trees, feet flying, crushing pine needles, startling birds.

He bashed on the cabin door with both fists. 'Come on out and fight!'

At the edge of the clearing, man-sized shadows moved into the trees.

The biggest boy, who they called Red, sneered at them from the open doorway. 'Look who it is! The babies!' He swung the door open wide to reveal the boys inside the cabin, who huddled together, talking in low, urgent voices.

Will caught the glint of a knife in Red's hand. 'Give them back!'

'Get on your belly and crawl,' Red jeered, and pushed him hard in the chest.

Will staggered back. Behind him someone taunted, 'Come out, yellow bellies!'

Will rushed forward, howling and punching the air, but Red shoved him with both hands, and Will sprawled in the dirt. As he got to his feet, Red slapped his face with an open palm. Will's cheek stung and his eyes filled with water. The boys behind Will shrank back.

'Crybaby!' Red laughed.

'Am not.' Will used his head as a battering ram and threw himself at Red's stomach. They both fell, rolling and grabbing at each other, punching and struggling. Some of the boys inside the cabin ran out. One threw a punch and it was game on: they shoved and spat at one another, their faces contorted with fear and rage. The air was filled with shrill, frightened cries.

Social psychologist Muzafer Sherif, disguised as the camp caretaker, scribbled excitedly in his notebook, hardly able to tear his eyes away from the boys rolling and punching and kicking. Here was the proof for the theory that he'd been working on for years, that normally upstanding and fair-minded eleven-year-olds could turn into brutal savages. An observer coming across the scene, he later wrote proudly, would never have known that these 'disturbed, vicious … wicked youngsters' were actually boys who were 'the "cream of the crop" in their communities'.

Will rolled on top of Red, but Red grabbed his hair and pulled hard. They were both howling, but neither would let go. Then hands were trying to pull them apart, and Will heard adult voices. Will tried to resist but the man had hold of his arm and pulled him to his feet. 'Cut it out now, fellas,' the man said. 'Or someone's gonna get hurt.'

Will was too busy trying to take a swing at Red to notice how pleased the man sounded.

# Part One

# 1
# Tangled Beginnings

It was a hot, overcast summer day in 1954 as a yellow bus drove twelve boys towards the foothills of the San Bois Mountains in south-eastern Oklahoma. The drive from Oklahoma City took just over four hours, and most of the boys, who had been strangers when they boarded the bus, had made new friends by the time they arrived.

Social psychologist Muzafer Sherif had chosen the location carefully. The attractions of the nearest town, McAlester, with its soda fountains and picture theatre, were far behind, but the boys didn't mind. The Robbers Cave State Park would be theirs for the next three weeks. It was their first summer camp, a sojourn that promised excitement and adventure. Outside of McAlester, the bus took a winding dirt road and ascended the mountain. Thunder rolled in the distance, promising rain that never came.

They turned off at a pyramid of logs carved with the letters 'Robbers Cave State Park', named because it was the hiding place of legendary outlaws such as Jesse James and Belle Starr. It seemed an ideal setting. A handful of pretty stone cabins dotted the treed park. There were lakes and a stream for swimming, a mess hall, a baseball pitch. And the mountains, with their hidden caves.

The boys explored the park, hiking, canoeing, and swimming. After they saw two rattlesnakes at the creek, they named themselves the Rattlers. For a few days they had the place to themselves, and pretty soon they felt as if they owned it. What they didn't know was another yellow bus had arrived, carrying a second group of boys from Oklahoma City, and that soon they would be locked in fierce competition.

Two weeks later, their faces covered with camouflage paint, the Rattler boys crept up in the darkness towards the cabin of their enemy, the interlopers who called themselves the Eagles. The Rattler boys raided the cabin, upturning beds, emptying suitcases, and terrorising the frightened Eagles. The midnight raid sparked days of retaliation and violence.

Nothing like the Robbers Cave experiment had been done before or has been done since. Dr Muzafer Sherif and his team of researchers, disguised as camp staff, closely observed the two groups during what the boys had been told was a regular summer camp. Sherif predicted that when he brought two groups together in a week-long contest of games and feats of skill, for which only one group would win a much-valued prize, the boys' attitudes would intensify from friendly rivalry to something closer to hatred. Normally well-adjusted boys would become 'nasty' and 'aggressive' towards the members of the rival group.

Social psychologist Muzafer Sherif is best known for this experiment in the wilds of Oklahoma, and what it says about the bonds of loyalty and the power of groups. He argued that it wasn't the boys' personalities that made them behave like savages

— after all, he had deliberately chosen normal, well-adjusted kids. And it wasn't differences in ethnicity or religion, either. He deliberately chose boys from similar backgrounds. Prejudice and conflict arises between groups of people, he argued, because of competition for limited resources. When groups of people compete for a valuable prize and there's only one winner, hatred and violence is inevitable. But it is reversible, according to Sherif. If groups cooperate to find a solution together, the boundaries between them dissolve, hatred is forgotten, and enemies become friends.

When Dr David Baker at the Archives of the History of American Psychology in Akron, Ohio, told me that the Sherif family had just donated all the papers, films, and recordings related to the Robbers Cave experiment, I only pretended to know which experiment he was talking about. As far as I recalled, Sherif's work was not included in my undergraduate psychology textbooks at La Trobe University or in any of my subsequent postgraduate training. That might have been because the kind of psychology I studied was very much the 'rats and stats' variety that aligned itself with the hard sciences. Bona fide social psychological research at my university had been laboratory-based and highly structured, using adult subjects and statistical techniques that reflected a kind of rigour and control and attested to the scientific credentials of the researcher in charge. A three-week field experiment that relied heavily on observation would have belonged over in the school of sociology, alongside the work of Sherif's mentors and colleagues, such as William Foote Whyte and his study of a Chicago slum, or in the school of anthropology, alongside Margaret Mead's equally famous research in Samoa. The truth is, Sherif's work didn't fit neatly into either discipline, and after I began to research his life I began to see this difficulty in categorisation as a metaphor for Sherif himself.

Dr Baker told me how the Robbers Cave would make a great book. But I was visiting the archives towards the end of four years of research about Stanley Milgram's obedience experiments, and the prospect of embarking on another project was the last thing on my mind. As part of my research for the book on Milgram, I had listened to hundreds of gruelling hours of recordings of the obedience experiments, in which participants believed they were delivering electric shocks to a stranger: the repeated cries of an actor pretending to receive electric shocks and the pleadings and distress of the men and women being ordered by a stern experimenter to keep increasing the voltage. I'd also tracked down and met with some of Milgram's subjects, who, fifty years later, were still agitated and troubled by the experience. In the course of it all, I had traded the idea of social psychology as an exciting discipline that shed light on human nature for a much more cynical view. Experiments such as Milgram's, that I'd so admired at university for their clever construction of elaborate theatrical scenarios to disguise their purpose, seemed to offer little beyond the rather obvious conclusion that people can be deceived and manipulated into doing things they would never normally do. As for the experimenters, I had spent enough time trawling through Milgram's papers to feel disenchanted with the valorisation of the daring and brilliant scientist whose fearless pursuit of the truth justified any pain he might have inflicted on his subjects. I was burnt out.

Sherif's Robbers Cave experiment might have sat at the intersection of the disciplines of anthropology, sociology, and psychology, but it was similar in some key ways to Milgram's obedience experiments. Both used deception and subterfuge to observe subjects in secret; both constructed a scenario where participants in the drama faced a moral dilemma. In Milgram's lab, the subjects' dilemma was whether to obey or disobey an experimenter's instructions to give electric shocks to another person. In Sherif's state park, the boys were caught between the dictates of conscience that told them not to inflict harm and the powerful pull of their need to belong. I felt angry on behalf on Milgram's subjects, whom I'd come to see as victims of a discipline that cloaked lessons about humanity's moral weakness and vulnerability in the language of science. I had no appetite for research into another experiment of that kind, especially given that Muzafer Sherif's subjects were children.

But I feigned interest as Dr Baker showed me round. I spent the rest of the afternoon looking through files and folders, taking photographs of documents, in a desultory and half-hearted way. Yet in the space of those few hours, I began to see that Sherif's work was markedly different to Milgram's. And the more I compared the two men, the more interested I became in Sherif.

In contrast to Stanley Milgram, who seemed to flit from one increasingly kooky experiment to another during his career, privately agonising over whether his research was more art than science — or if it meant anything at all — Sherif's papers revealed a man with a singular focus and an apparently unshakeable faith in his own theory. Sherif made Milgram look like a dilettante. It wasn't just the volume of material Sherif wrote that struck me, but the fact that over a fifty-year research career, he explored variations of a single theme: the power of tribal loyalty, in-groups and out-groups, to shape our worlds.

Both Milgram and Sherif devised experiments that were part of a wider struggle to understand how ordinary people came to participate and collude in the brutality and violence of war. For Milgram, exploring the behaviour of Germans under the Nazi regime, the question was whether it had something to do with the German character. Had the surge of anti-Semitism encouraged by National Socialism been lying dormant all along? Milgram attributed it to a universal instinct for following orders. Sherif was convinced that the answer lay in the power of the groups we belong to and identify with, which shapes the way we behave. For Sherif, a group was more than a collection of individuals: once bonds were formed, tribes developed their own culture, with leaders and followers, rules and standards distinct from the ones an individual might hold dear. The group takes on a life and a personality of its own.

Milgram's shocking results, as well as an entrepreneurial streak and gift for self-promotion, propelled him and his obedience experiments into lasting fame. But when I googled Muzafer Sherif and the Robbers Cave experiment, I was surprised at how brief both pages on Wikipedia were, offering few details beyond a summary of the research and the bare facts of Muzafer Sherif's life. There were articles about the experiments, of course, and academic monographs. But still. Why wasn't Muzafer Sherif better known?

He certainly had the personality and the chutzpah. Many of his contemporaries describe him as a man who loved being the centre of attention, who could be charming sometimes, arrogant others, and was always convinced of his own genius. Sherif took credit for converting shy young Solomon Asch to the idea of studying psychology during their time at Columbia University. The social psychology family tree being what it was, without Solomon Asch there would be no Stanley Milgram. Sherif bragged to colleagues that with the Robbers Cave experiment he'd broken the mould of social psychological research. Yet Sherif made only one foray into the popular press, in 1969, fifteen years after the Robbers Cave experiment was over, when he wrote an article about it for *The Washington Post.* A Hollywood producer contacted him almost immediately, saying he wanted to make a film based on the experiment, but Sherif refused to have anything to do with it. If Sherif's reticence wasn't due to shyness, I wondered, what was it that made him shrink from the public gaze?

The small amount I knew about Muzafer Sherif was contradictory and confusing — he both craved attention and shied away from it, he was cautious and a risk-taker, highly regarded and an outsider.

The brief biography online told me that he was born in Turkey and his American career started with great promise, first at Harvard, then at Princeton and Yale. So how and why did Sherif end up at a comparatively lowly university like the University of Oklahoma? What led him to these remote locations, these groups of warring boys? If, as historian Jill Lepore argues, psychological research is autobiographical, what mirror did the Robbers Cave experiments hold up to Sherif's own life?

Archivist Lizette Royer Barton pulled on white cotton gloves to carefully unpack a grubby calico flag. It was the size of a couple of pillowcases, featuring a childish painting of an eagle with a snake in its talons. The flag was among hundreds of items that Sherif brought away with him from the experiment and stored in a wooden trunk that eventually made its way to the archives after his death. Lizette then carefully unfurled a paint-spattered pair of jeans with the words 'The Last of the Eagles' painted in orange capital letters down both legs. In photos of the experiment, you can see a group of boys holding them on a pole like a flag, taunting their opponents. The theft and vandalism were acts of aggression, the denim trophy the symbol of a nation state in Sherif's eyes.

Looking at the jeans, with their narrow waistband, lying flat on the table, I couldn't help but think of the boy who had worn them. It was as if the boy himself lay on the table. They seemed further evidence of a branch of science that I had begun to think of as careless in its treatment of people, that used manipulation and deception to make a point and viewed the distress of human subjects as a necessary price to pay.

Later that day, in reading about how the Robbers Cave experiment came about, I found that Sherif had conducted two earlier similar experiments. In the first experiment, in 1949, Sherif took twenty-four underprivileged boys from New Haven to a remote farm called Happy Valley, in Colebrook, a town in the Litchfield Hills in Connecticut. After a few days where the boys mixed and played, he divided the friends into two groups and organised a three-day contest of games. Fighting broke out between the two groups after the winners were announced, and the prized knives went to the victors. Over lunch on the last day, the two groups were 'lined up on opposite sides of the mess hall calling names and finally throwing food, cups, table knives …'

Having proved his theory that friends will become enemies when they are forced to compete, Sherif declared the experiment over and instructed staff to 'do away with the hostility'. But it was easier said than done. Despite staff attempts to restore trust between them, the boys continued to retaliate against one another with fights, night raids, and surprise attacks. This gave Sherif an idea. In his second experiment, he looked at ways to reverse mistrust, hostility, and violence between warring groups and engineer a lasting peace.

While the first experiment was a success, his larger second one in 1953 had been aborted. I felt a tug of interest as I read this. What did a failed experiment about the inevitability of war and the restoration of peace look like? Had the boys resisted the attempts at peace or at war? It was difficult to find a straightforward answer. Sherif had published a number of detailed descriptions of his first experiment in 1949, and a whole book about his Robbers Cave experiment. But I could find only occasional passing references to the 1953 study, and almost nothing about why it had been cancelled. It was as if Sherif wanted to forget it.

But in the archives, unpublished material about this 1953 camp experiment took up almost as much space as the files and folders about his Robbers Cave experiment, held just a year later. There I found audio recordings, film footage, photos, detailed observers' notes, invoices, instructions, and letters about the 1953 study, all jumbled in together in seemingly little order. But there was no overall narrative to tie this mountain of raw data into a story of what had happened or how and why it ended.

Sorting through all that archival material and piecing the narrative together would be like trying to reconstruct a document from a mountain of shredded paper. But I already felt sure the story of the failed experiment was there in those papers and recordings, and I had the feeling that the answer to the contradictory and intriguing figure of Muzafer Sherif was somewhere in those boxes, too.

# 2
# In the Wild

On the other side of the Atlantic, just as Muzafer Sherif was wrapping up his first summer camp and preparing for his second, a schoolmaster was conducting a similar experiment high above England's Salisbury Plain.

On the road to Stonehenge there's a small brown sign pointing to a place called Figsbury Ring. The rutted and bumpy lane from the A30 climbs past a farm selling local honey and ends in a small, pot-holed car park. A final rise brings you out on top of a hill, where two concentric trenches trace the outline of an Iron Age fort and its ramparts. Up here, from the top of the mound, there's spectacular views of the Salisbury Plain, the spires of the cathedral glinting in the distance on a sunny day. It was here around 1951 that local schoolteacher William Golding, who was interested in understanding 'the nature of small boys', brought a group of schoolboys to this remote and ancient place. He divided them into two groups and instructed one group to attack and the other to defend the mound, looking to see what would happen if no adult intervened.

According to Golding, he got more than he bargained for. The two groups attacked each other with increasing ferocity the further their teacher moved away. 'My eyes came out like organ stops as I watched what was happening,' he told an audience in a lecture at the University of California in 1961. He did not say precisely what he saw, but he hinted that there was serious violence, even the risk of boys being killed: '"Give me liberty, or give me death" — well, it was a point where these were no longer simple alternatives.' It's impossible to know whether Golding exaggerated the violence in this episode, but it's no coincidence that artists and social scientists continents apart were conducting experiments on groups of children in conflict. The difference was that Golding's was carried out as research for his next book: the novel *Lord of the Flies*.

The reverberations of World War II and its mob violence drew both the writer and the social psychologist to the idea of warring groups of children. But their conclusions couldn't have been more different. Golding, who had served in the Royal Navy and taken part in the invasion of Normandy, returned home with a fervent belief that 'man produces evil as a bee produces honey'. He described the theme of his allegorical novel about a group of shipwrecked boys as 'an attempt to trace the defects in society back to the defects of human nature. The moral is that the shape of a society must depend on the ethical nature of the individual and not on any political system.'

Golding's view of human nature had more in common with mainstream psychology at that time than Muzafer Sherif's did. It was a variation on the riff of original sin — his belief in which, he said, came directly from his experiments, practical experiences, and observations. In his novel, Golding projected his view of the 'true' nature of savagery onto the children.

In Sherif's view, human nature had little to do with it. People were inherently good, and it was an environment — whether economic, political, or social — that fostered inequality between social groups and created the ideal conditions for discrimination and mistreatment to flourish, spilling over into conflict, violence, and war. An idealist, Sherif believed that by understanding group psychology, social scientists could work to dissolve prejudice and persecution, reduce nationalistic antagonisms, and foster peaceful co-existence across the world.

But the popular perception of an experiment involving children is that it captures something more raw and elemental. And like Golding's *Lord of the Flies*, the fact that Sherif's experiments took place in the wild, away from the distractions of city life, reinforced the idea that he had accessed something intrinsic to human nature. Despite Sherif's conclusion that hate is learned, his research is often twinned with Golding's novel, as if he too observed a 'natural' savagery of the kind depicted in *Lord of the Flies*.

Writers and social scientists such as Golding and Sherif, horrified and sickened by war, were not alone in regarding children as a canvas onto which they could project anxieties and hopes, fears and dreams. In the 1930s, research with groups of children focused on understanding the lure of Nazism and inoculating American society against it. In an experiment conducted late in the decade, German émigré psychologist Kurt Lewin explored how dominant political ideologies shape behaviour, using eleven-year-old boys as subjects. The boys took part in a craft workshop led by either an autocratic or a democratic leader. The results showed that boys whose leader was domineering and autocratic were more hostile and aggressive, while those with a democratically-minded and inclusive leader were more cooperative, friendly, and supportive of one another. Lewin's research pointed not only to the impact of the political doctrines of Germany and the United States but also suggested ways in which future citizens could be protected from fascism and moulded to embrace democratic ideals. For Lewin, the groups of children were both a template for new citizens and proxies for society at large.

A decade later, by 1949, funding of social psychological research had expanded to reflect the Cold War anxieties of the era. With the looming danger of an atomic attack, the spectre of totalitarian brainwashing, and the threat of communism, the US military became increasingly interested in the psychology of small groups. Kurt Lewin's students continued his work, studying groups of children in natural settings, superimposing the political anxieties of the era onto children at play. A group who made a game of pretending to sink to the bottom of a swimming pool was a metaphor for the stirrings of a collective mind; a group led to sing the same lewd song was an insight into how rumours and panic spread, into how a leader could either inspire or quell loyalty. This kind of play was labelled 'contagious', as if it was a sickness to be caught, a disease that would leap from person to person with frightening efficiency, creating a new and dangerous strain that could spread in ever-widening circles. The children were stand-ins for adults; the swimming pool and the campfire ring, for society at large.

Muzafer Sherif never publicly discussed or recorded in his

notes why he chose to use eleven-year-olds as his subjects, but he admired Lewin, and selecting subjects of the same age would have allowed for direct comparison between his and Lewin's work. Studying children had particular benefits, too. They were easier to deceive, so their reactions were more likely to be spontaneous and revealing. And ten- and eleven-year-olds had a level of maturity and social skills that younger children lacked, but hadn't yet developed the adolescent rebelliousness that could make them uncooperative.

Some exasperated researchers yearned for a research environment that allowed continuous access to child subjects, free from the interference of home life and other 'contaminants' that could undermine the purpose of their research. Summer camps provided just such a setting. Child psychologist Mary Northway wrote in 1940:

> The summer camp offers an ideal field for research for the social psychologist. It is an isolated, constant, temporary group, as far removed from the ordinary roads of social intercourse as a south sea island. Camp suddenly comes into existence when a group of individuals, cut off from the ties of their normal societies, are thrust together in one geographic community, and a new society is created. While camp lasts, it is an isolated community; and it may be considered a society in miniature.

By the time Sherif began his first study in 1949, using summer camps as psychological laboratories was a well-established practice. But what he had in mind was a new twist.

The first organised summer camp in North America had a character-building focus, and camps ever since have continued this tradition — although definitions of good character have changed with the times.

Early founders of the camping movement were educated men worried about the evils of progress. The original American camp took place in 1861, when schoolmaster Fred Gunn took his pupils on a two-week stint of outdoor activities to improve their mental and physical development under the guidance of adult mentors.

Just under twenty years later, in 1880, Camp Chocorua was established by tutor Ernest Balch to toughen up wealthy boys whom he considered at risk of effeminacy because they spent too much time with their mothers. Balch had plenty of marketing savvy, and his view that modern young men risked growing soft in the lap of luxury, doing nothing and learning nothing of value during the indolent summer months, was covered widely in popular magazines such as *McClure's* and *Outdoor Life*. Camp leaders and parents embraced the idea that lessons in masculinity could be learned from a structured outdoor life — the same lessons learned by the pioneers, whose contact with nature made them strong and hardy. The mainly urban families who signed up their sons for these camps hoped the boys would develop physical sturdiness, moral character, self-reliance, resourcefulness, and Christian values.

Initially, camps based on the back-to-nature movement were reserved for the wealthy. By the turn of the twentieth century, social reform and charitable organisations such as the Fresh Air Fund were running camps for poor city children to give them a break from urban life and a chance to build up their physical strength. In many cases, the children were also offered a dose of religious instruction. 'Feed 'em well and let them breathe fresh air' was the motto. Daily camp life followed a highly structured routine, beginning with the wake-up bell or bugle call, breakfast, chores, and days filled with athletic activities, games, and team competitions.

Summer camps adopting similar principles flourished across North America between the late 1880s and early 1900s, and were given fresh impetus with the arrival of the Boy Scout movement in 1910. Robert Baden-Powell, a British army officer, attributed the disastrous performance of his army during the Boer War — where British victory was achieved only after huge cost and massive reinforcements — to weakness of character, demonstrated in soldiers' poor attitudes towards authority and inexperience in living outdoors. He conducted his first two-week 'boy training' camp in 1907, housing twenty-two boys on isolated Brownsea Island, in England. The idea quickly spread: before 1900, there were less than a hundred camps in North America, but by 1918 there were more than a thousand.

The newly emerging discipline of psychology added a new dimension to the romanticised idea of taking children back to nature as an antidote to the perils of urban life. Popular psychologists such as G. Stanley Hall endorsed summer camps as places where children could benefit psychologically from an escape to a simpler and less complicated world.

After World War I, camp leaders influenced by the Progressive social theory of John Dewey began to design camp communities that mirrored the adult communities they wanted to create. Progressives viewed the camp experience as a means to strengthen democracy, build a more just and equitable society, fight authoritarianism, and foster the principles of good citizenship. In 1927, Professor Elbert Fretwell began the first camp leadership course at Columbia University, and a new profession was founded. Camp administrators drew on psychological research for the latest information on the best ways of working with groups.

Progressive educators took a scientific approach towards child behaviour and educational theory. So it didn't take long for psychologists to get involved in evaluating the best techniques for shaping camper behaviour. Camp leaders began using scientific instruments and tools such as sociograms — diagrams of how campers felt about one another — to identify boys who weren't popular and take corrective action. Psychologist Richard Doty applauded psychological tools that could both diagnose and treat problems: 'We can quickly set to work with the less accepted campers to develop skills and attitudes making for better group acceptance and more positive and acceptable social philosophy.'

Therapeutic camps for youth with mental or behavioural problems opened up in the 1920s. The first was Camp Ramapo, in upstate New York, for 'delinquent and problem children'. Apart from camp counsellors, psychiatrists and psychiatric social workers were also part of the staff, conducting observations, making diagnoses, and recommending further treatments. In a kayak or in a bunkhouse, in the dining hall or on the baseball pitch, the agenda of summer camp was to teach problem children the social skills for interacting successfully with peers. Away from the formal environment of the school or the clinic, in the relaxed atmosphere of the great outdoors, camp staff offered one-on-one talk sessions and interventions to help children overcome phobias and fears, build self-esteem, deal more assertively with bullies, or learn to control their aggressive or delinquent behaviour. Kurt Lewin's famous studies on the effect of democratic and autocratic leadership styles on groups of children were incorporated into staff training to demonstrate the powerful effects of the camp counsellors' style on camper behaviour. The idea of the summer camp as a place to transform the individual was now not just linked with theology or social reform, but also with science.

Yet notions of what constituted the ideal camper diversified too. In the 1920s and 1930s, the ideological function of summer camps expanded to accommodate broader political perspectives. 'Red diaper' camps sprang up for the children of more politically radical parents, as did Hitler Youth camps for the children of the German American Bund. In 1938, film footage of young American boys dressed in Nazi-inspired uniforms doing calisthenics, shooting rifles, and saluting the Nazi flag caused a public scandal. The relatively benign view of summer camp as a place for shaping character and turning out model citizens had taken on the sinister cast of indoctrination and brainwashing. The US government responded by setting up a committee to investigate the subversive activities of the German Bund. When it had finished, it turned its attention to the activities of American communists under the committee's new head, one Senator Joseph McCarthy.

In the late 1940s, with the beginnings of the Cold War, the US government stepped up its efforts to develop weapons of psychological warfare, funding a network of communication studies centres across the country to continue the research efforts into propaganda that were established during the war. One of them was Yale's Communication and Attitude Change Program, where Muzafer Sherif was working. The program's neutral-sounding name deliberately avoided any association with the negative or manipulative connotations of propaganda, but the link to psychological coercion was there in its research. Programs such as Yale's benefitted from the government's obsession with psychological warfare, and social psychologists gained from this bonanza of government funding to study techniques of persuasion that could be used to deal with international conflict abroad as well as control of society at home. But by the late 1940s, America was involved in a new war, one that was as much about two competing ideologies, and the propaganda battle shifted to finding psychological weapons to fight the spread of subversive ideas.

Sherif's mentor at Yale was Carl Hovland. One of Hovland's research interests, gleaned from his war-time study of soldiers reluctant to fight, was the power of small groups in changing attitudes and behaviour. When it came to propaganda, Hovland found that fellow soldiers in the same troop were far more effective than military films at convincing unwilling soldiers to enter battle. Sherif's idea for an experiment on the power of groups to shape the attitudes and prejudices of its members seemed, on the face of it, to continue in a similar vein.

Knowing Sherif's interest in group influence, in March 1948 Ronald Lippitt, a former student of Kurt Lewin who had visited Yale to give a guest colloquium that year, invited him to collaborate in his research on children at summer camp, but Sherif turned him down. Observing children in a single group, as Lippitt had planned, was of no interest to him. But Lippitt's study gave Sherif an idea. What better way to study the interactions between groups of boys than by running his own summer camp?

In September 1948, Sherif approached the American Jewish Committee, who published research into anti-Semitism, for $5,000 for his first study. In December, the chairman wrote that they did not normally fund research but they were bowled over by his enthusiasm: 'One of the chief reasons for my interest in this project … is my realisation of its great meaning for you. Research which has poured into it a considerable amount of the researcher's libido … turns out to be much more significant than research which is done for other reasons.'

Sherif wrote back gratefully: '[T]his particular research plan means an awful lot to me … [if it] will further just a little bit our understanding of group frictions and thereby eventually contribute one whit to their elimination, this will be the greatest reward for me.'

But he must have had his doubts about whether he'd be able to pull off what he had in mind. His plan was audacious and daring.

Among Sherif's papers there's a copy of the training manual for a four-day course for summer-camp staff. Titled *Camping with Children, or Living and Learning Democratic Human Relations Together*, the twenty-five-page handbook describes in detail what makes a good camp counsellor, and was a blueprint for staff training at the time. The ideal camp counsellor was someone well-versed in child and group psychology; who could foster cooperation and build strong relationships between the children; who was a good role model, an inspiring leader, a skilful problem-solver, and an empathic listener who offered counselling and guidance. The counsellor was responsible for the 'happiness of their campers', and happy campers were those who paid attention to and were responsible for other campers and staff.

The manual presents an idealistic and humanistic vision of summer camp as a place that develops children to be good leaders and responsible citizens who care for others, and ensures the greatest benefit for the group by developing the skills to live and work in harmony. But the experimental summer camp Sherif had in mind would require camp staff to do the opposite of their training: to abandon ideas of fair play and democracy and swap their commitment to the happiness of each boy in their care for the pursuit of scientific research.

While the camp counsellors were running a standard summer camp, with its usual activities — swimming, archery, hiking, sing-a-longs around the campfire — there would be a second, behind-the-scenes camp to conduct. In this shadow camp, experimenters would pull strings to ensure not a single happy group of campers but two hostile groups who viewed each other with suspicion. One layer of the camp would be the world as it should be — democratic, harmonious, fair — and the other would be the world perhaps as Sherif saw it — unjust, divided, full of violence and conflict.

For Sherif, finding staff who would embrace an upside-down summer camp, one that inverted an American institution and encouraged all the behaviours standard camps aimed to eradicate, would be no small job. And how would he gain the approval of parents to entrust their children into his care?

Sherif's letters to parents are a lesson in the art of skilful deception and subtle persuasion. And he got better at them as time went on. Taking a leaf out of his mentor Carl Hovland's book, Sherif used trusted sources in the community to allay suspicion and persuade parents of the benefits of allowing their boys to participate. In his letters, he appealed to the parents' interests and provided just enough information to get their consent without alarming them about what might be in store.

For his first experiment in 1949, Sherif asked Episcopalian ministers in New Haven to identify and give him the addresses of poor and underprivileged families with eleven-year-old sons. He reasoned that such families were more likely to be swayed by the offer of a free camp. His subjects needed to be of 'normal or higher intelligence, with no physical defects or serious emotional problems'. Sherif reassured ministers that religious observances would be maintained at the camp, with Sunday services and staff leading boys in saying grace before meals. Sherif was confident: he wrote in notes beforehand that, given the camp was free and included food, 'it will be easy to get boys'.

In his letter to parents, he was vague on details: 'The Yale Department of Psychology is co-operating in a study of child relations and social organization among children', he wrote. A condition of agreement was for parents 'not to visit' because it might 'distract' the boys. In a draft of his final report on the experiment, he later wrote with satisfaction, 'The parents and the boys were told that new methods of camping were being tried out. They never suspected there were any psychologists in the camp. They did not know that any manipulation of conditions for experimental purposes would be done or observations of behavior made.' However, Sherif was bothered that some boys in the study were loners who had been reluctant to join in group activities.

For his second experiment in 1953, Sherif recruited boys from in and around Schenectady, New York. When he contacted local ministers, he was more specific about who he was looking for: 'middle class … normal, healthy Protestant boys' aged ten-and-a-half to twelve-and-a-half who were 'well adjusted … "typical American Type"[s] who were "group minded" and who came from well-adjusted families, not from "broken homes"'. He didn't want boys with signs of 'delinquency or involved emotional problems such as frustration, strong mother attachment, etc.' He didn't want boys who were 'cissies' or 'social isolates', or boys who were '*primarily* interested in solitary recreational pursuits and hobbies: fishing, insect collecting, stamp collecting, etc.'

His letter to the parents was different, too:

> For many years camp executives throughout the country have been trying to find out what camp activities will result in giving their campers a fruitful educational and recreational experience. These camp directors are interested in finding out what things can be done to give their boys and girls a wholesome cooperative living experience which will prepare the youngsters for better citizenship and to be leaders in their communities. The question is what camp programs best serve to enrich the life and experiences of growing children?

The purpose of this camp was 'simply to study the best programs and procedures for campers which will develop cooperative and spiritual living'.

Instead of appealing to parents' pockets, this time Sherif appealed to their aspirations for their children. The boys would be 'carefully selected'; successful boys would receive a 'scholarship'. He emphasised that the study was auspiced by prestigious institutions, including Yale and Union College. His letter portrays the camp as benign and instructive, and selection as a camper a privilege. There was no indication that parents were volunteering their children for a three-week psychological experiment that would require them to navigate some dubious moral terrain.

Fat, thin, quick-witted, slow, tall, short, outgoing, shy — William Golding would have identified each boy in his classroom by his distinguishing characteristics. But on the ancient hilltop at Figsbury Ring, their individuality was lost, and they merged into warring tribes.

Muzafer Sherif was not nearly as interested in individual boys as he was in groups. Already in the design of the recruitment letters he sent to ministers and parents, his focus was on their similarities, rather than their differences. In all three camp studies, he chose boys who were 'homogenous', their individuality swallowed up in generic categories of age, religion, hobbies, and family background.

By now I was gripped by the story of the experiments and had decided to write about them, but the characters were slippery. Encountering the experiments for the first time myself, it was difficult to keep track of individual boys — especially since in retellings of the experiments, Sherif and subsequent generations of social psychologists have often conflated his three camp studies so that the groups become even more blurred. And while this sense of anonymity, of universality, adds to the power of Sherif's conclusions, it allows us to forget these groups of subjects were made up of individuals, each with a particular history and personality. What did they each make of the adult observers and their watchful distance from activities that would usually bring camp counsellors hurrying to intervene? When did they each notice that the usual rules for summer camp did not apply, and how did they feel about this?

By imagining them as a crowd, a mob, a mass, I felt complicit in a process of forgetting and erasure. But this is the dilemma of social psychological research: that the human subject becomes an object representative of a world of the social psychologist's imagining. In the process of gaining insight into human social life, the researcher loses sight of what it is that makes us human — those qualities that define us as separate and unique.

In telling the story of his research, Sherif had plenty of material, in the form of observation notes, photographs, and even some film footage, much of which he never used in his final published reports. Of course, a process of selection is inevitable when social scientists are writing up their research, as they sift, choose, and shape the material they have into a story. But what's often fascinating is what they leave out.

It seemed to me that in order to tell a balanced story, one that gave equal weight to the perspective of the researcher and the researched, scientist and subject, I had to re-create the world of Sherif's summer camp and, in a story until now narrated by social psychologists and professors, make space for the voices and memories of the boys themselves.

# 3
# Lost and Found

Perhaps science's obsession with the notion of the lost boy began with nineteenth-century studies of feral children and the insights they offered into human nature. Perhaps psychology's obsession started with Little Albert — or maybe it started long before him.

In the winter of 1919–1920, eleven-month Albert was a rosy-cheeked blond baby whose mother worked at the hospital at Johns Hopkins University in Baltimore, Maryland, where the charismatic Doctor John Watson, regarded as the father of Behaviorism, had set up an infant laboratory. You can see pictures of Little Albert online, leaning away from Watson, who is looming over him, wearing a scary mask. Little Albert is famous as the child subject who proved Watson's theory that emotional responses can be learned and unlearned, a process that became known as classical conditioning. Watson paired the loud and frightening noise of a hammer hitting a metal bar with the appearance of a small rat. Albert, who had reached happily for the animal at first, soon came to associate it with the terrifying noise and became agitated and afraid whenever the rat — and, later, other furry animals — appeared. We'll never know whether Albert's mother volunteered Albert willingly and discovered too late what the research involved, or whether she was coerced to hand Albert over. But she took Albert away abruptly, before Watson and his assistant, graduate student Rosalie Rayner, had a chance to de-condition the baby's fear of furry animals.

Soon afterwards, Watson, the psychologist superstar, was sacked from his position at Johns Hopkins University for having an extramarital affair with the twenty-one-year-old Rayner. The scandal made front-page news across the country, and the pair fled to Manhattan. After a shaky start, Watson enjoyed a successful second career as an advertising executive.

The fate of Little Albert was unlikely to be so lucky, according to Watson and Rayner. Historian of psychology Ben Harris noted that although at first they said that any fear he experienced during the experiments was transitory, in a magazine article in 1920 they wrote that Little Albert 'had probably suffered permanent harm', a claim made not from callousness so much as to emphasise the power of the conditioning they had conducted on the baby. So the mystery of what happened to Little Albert and how he was affected by the experiment remained.

In his review of variations of the Little Albert story in textbooks, Harris points to significant deviations in Watson's own reporting of the research, and writes about how startled he was to read the original article in which Watson and Rayner commented on their lack of success in inducing fear in the rosy-cheeked baby. The story has evolved over time, embroidered and shaped to support a particular theoretical standpoint and to make a comment about psychology in general — and, as Harris points out, to portray Watson as a martyred hero in the development of the discipline. 'He undertakes work that no one else would,' Harris writes of the perception of Watson that took hold. 'He unemotionally pursues the truth while surrounded by the superstitious. Unfortunately, his perfectly designed experiment on Albert was cut short by the infant's ignorant, ungrateful mother. Blackmailed by his wife, betrayed by Albert's mother, Watson becomes a tragic figure, a victim of his single-minded pursuit of science.'

If Watson was portrayed as the tragic star of this famous story, what of Little Albert? Historians of psychology have been on the trail of the now elderly or deceased Albert. Psychological sleuths Hall P. Beck, Sharman Levinson, and Gary Irons were the first to claim that they'd found 'psychology's lost boy'. Albert, they wrote in 2009, was a child called Douglas Merritte who had died aged six of hydrocephalus. His mother had worked as a wet nurse at the orphanage opposite the hospital where Watson and Rayner were based and volunteered her baby for the research.

Such was Little Albert's fame by the time of the researchers' discovery that the BBC sent a film crew to record them visiting Merritte's grave. Their find elicited huge public sympathy for the child. One reader commented on a blog that featured the story, 'Little Albert is and will stay the James Dean of psychology, an experimental icon.'

In an update on the story in 2012, psychology professor Alan Fridlund and his co-authors argued that Watson knew during the experiment that baby Douglas had a neurological condition with symptoms that affected his motor and social skills and would have influenced his behavioural conditioning. They argued that in failing to disclose that his results were based on the behaviour of a child who was gravely ill, Watson had engaged not only in a serious breach of experimental ethics but also perpetuated a case of academic fraud.

The story was widely reported, and by 2014 a new myth was born, with psychology textbooks amending their portrayal of both Little Albert and Watson — the child portrayed as 'neurologically impaired' and the experimenter as 'recklessly unethical'. Even authors of non-academic psychology books incorporated the story; for example, Joel Levy's *Freudian Slips: all the psychology you need to know* points out that '… since Little Albert was not a healthy child, whatever value the study may have is destroyed, along with the remains of Watson's reputation'.

But the Little Albert story was not over. In 2014, with the help of a professional genealogist, and after close examination of Watson's film footage, psychologist Russell Powell and his colleagues concluded that Douglas Merrite was not Little Albert: they had found a more likely candidate. Little Albert, they argued, was a healthy child called Albert Barger. Quite apart from rescuing Watson's reputation, here was an opportunity to put Watson's claims for the power of learned responses in childhood to the test. Was adult Albert still frightened of furry animals?

Sadly, the researchers couldn't ask Albert himself because he had died in 2007, at the age of eighty-seven. Instead they turned to the recollections of Albert's niece, who had been close to her uncle. The results were tantalisingly inconclusive. She reported that throughout his life Albert's family had teased him about his aversion to dogs, and recounted how dogs had to be kept in another room if Uncle Albert was visiting. Was this, as Albert had explained to the family, the after-effects of seeing his pet dog killed by a car when he was a young child? Or was it a legacy of Watson and Rayner's experiment? Either way, the researchers concluded that Albert had never been told of his role in Watson's research.

If Little Albert was psychology's original 'lost boy', it seemed to me that Sherif's subjects — from all three studies — were in some sense his descendants, and his compatriots: the lost boys. No one had ever set out to find them; no one had ever tracked them down. Like most of the subjects in psychology experiments in the first half of the twentieth century, they were nameless, faceless individuals who disappeared back into ordinary life once the experiment was over. Would finding them provide any definitive conclusions about the power of Sherif's experiment? Would they carry some hidden legacy from the research they took part in as children? And what message had they each taken away from the experience, about themselves or about other people? Would they have grown up as warmongers, or peacemakers, or did the experiment have no effect on them at all? And what were their individual stories? Would finding the boys from the 'lost' experiment in 1953, with its hints of mutiny or resistance, provide a missing piece in the story of Robbers Cave?

'I'm not traumatised by the experiment, but I don't like lakes, camps, cabins, or tents,' Doug Griset told me. 'My kids always said, "Why is it, Dad, that you never want to go camping?"' Doug laughed. 'I always told them about this camp where I ended up in hospital, but no one believed me.'

We were talking on the phone, and he sounded excited. Doug

was the first boy I found from the abandoned 1953 study. It was our first phone conversation, after a flurry of emails, and in speaking to him I realised that he had never been told that the camp was an experiment. I hadn't expected to lob such a grenade into someone's life, and after my experience with Milgram's participants, I understood how troubling this revelation could be.

'It's funny,' Doug said. 'But my wife, June, and I have been addicted to this TV series called *The Fringe*, about a group of young people who find out when they are grown up that they were experimented on as children. What do you make of that?'

What did I make of it? It never occurred to me that the children in Sherif's research hadn't been told later that they were being studied, and I would be unexpectedly in the situation of having to explain the experiment to Doug and tell him something about the man behind it. I thought of the half-dozen other letters I had sent out across America introducing myself, asking people if they were the same person who attended a camp near Schenectady in upstate New York in 1953. It must have been strange enough to receive a letter like that out of the blue, but it was weirder still to have someone tell you that sixty years ago you were experimented on.

Yet Doug was fascinated. Suddenly his life had a strange twist, an interesting plot point. He told me the end of the story first. 'I remember my parents coming to get me from a hospital in the Adirondacks. I think it was in a place called Monroe, close to the Canadian border. Would that be right?'

Our roles were reversed. Doug was asking me questions instead of answering mine. So between our initial phone contact, in June 2011, and our first meeting, almost a year later, I went back to that archival material and tried to piece together as much as I could about the 1953 study. But even its exact location was a mystery.

The trouble was that it was a study Sherif clearly preferred to forget. He made reference to the failure of the experiment in the footnote of an early edition of one of his books, and revised this footnote in a later version. Both instances were worded awkwardly, but each told a different story about the failure of the experiment. A 1954 footnote read that the experiment was aborted because of '… various difficulties and unfavorable conditions, including errors of judgment in the directing of the experiment'. Two years later, he wrote that 'in a frustration episode, the subjects attributed the plan to the camp administration'. It was hard to decipher just what he meant in both footnotes. Was the language deliberately unclear? Which was it: the experimenter's fault or the boys'? I had expected Doug would be able to tell me, but after speaking with him I realised I would have to look further for the answers.

When we met, I showed Doug a copy of the letter Sherif wrote to parents such as his, emblazoned with the monogram of Union College, a prestigious institution that lay claim to being the first non-denominational college in the country. With its suggestion that this was a personal invitation to their son, parents like Doug's (and especially his mother, according to Doug) may have hoped that this letter would trigger a flow of other invitations — to university gatherings, tennis parties, bridge nights, perhaps even membership of the golf club.

Sixty years later, Doug laughed and shook his head when he read the letter. He sat in the lounge room of his home, not far from Union College, on a nineteenth-century housing estate known as The Plot. The college had originally owned this 30-acre tract of wooded land. In 1898, they sold it for $57,000 to the General Electric Company, who wanted to build a housing estate for its employees. But these weren't just any homes. And these weren't ordinary employees. From 1898 to 1927, the company built around one hundred enormous homes in an eclectic mix of styles that it hoped would lure the best research scientists from around the world. Swiss chalets, Tudor homes, Queen Anne–style mansions, and Spanish colonial houses sit back from the street on deep, large blocks, shaded by established gardens. It is said that this geographic patch generated more scientific inventions than just about any other patch in history, as General Electric employees continued their scientific tinkering in their homes and sheds at night, after they got home from work. We're sitting in a house where the first TV broadcast was received, in 1927.

Science was the lifeblood of Schenectady. In 1951, when Doug and his family moved here, General Electric and Alco, both renowned for their scientific discoveries and contribution to the war effort, were the largest employers in town. Today, the large silver letters *General Electric* dominate the skyline above the imposing red-brick headquarters on one of the town's main roads, although the number of employees is dwindling.

Doug peered at the letter again. 'That letter was the perfect sell!' he said. 'Yale, Union College.' He laughed and slapped his knee. 'The only thing missing is God and America. What an amazing — I mean, somebody ingenious drafted that letter.' He passed it across to June.

June started reading the letter half aloud, but Doug kept interrupting. 'Don't you love it? Just picture my mother and father getting this.'

'A grant to Yale, that would have been impressive.' June's lips twitched. 'Oh, the Rockefeller Foundation —' she said with a laugh.

'The Rockefellers! Can you see my mother? My twenty-eight-year-old mother reading this? It may as well have come from the President of the United States!'

'The Rockefeller Foundation,' June said again.

'Yeah, from Yale! They mention everything except God and country. Oh,' Doug shook his head, still laughing, 'that would have got my mother. She was a sucker for anything fancy. My dad, too. All you had to say was "Union College" in those days. We weren't allowed on those premises — I never set foot on Union College and yet my father used to revere Union College. Oh, if you were a Union College person, whoa! That was a big deal. So imagine throwing Yale on top of that.'

'Looks like they had to do some kind of interview,' June said, scanning to the end.

'Pfft, not a problem,' Doug scoffed. 'My mother grew up dirt poor in the South, but she was smart enough. And my father could have finagled anybody he wanted. My father would have walked into any interview and pulled it off. So they would have handled an interview very well.' He thought about it for a minute. 'My dad, who was an extremely patriotic World War II veteran, could have been talked into sending me to something very easily — even though he was a very bright and discerning man — if he thought it was going to be advancing some patriotic concept. *We're gonna help figure out how we can make kids like your son be leaders*: that would have appealed to his sense of patriotism.

My mother, who came from a very poor background, would have been persuaded if she thought that she or the family would somehow be elevated by my attendance. But my parents were very protective. My mother — I wasn't leaving her side unless it was for something that she thought was wonderful for me, and it would have been pretty much the same for my father, except his view would have been, *Okay, you can go if it's gonna toughen you up a little bit.* Until I was about seventeen I was always the smallest kid in my class.

'But here's the part that still is gonna trouble me and trouble me. It's three weeks — that's a long time for a kid to be in a situation like that. My mother — that would have been a very long time for her too.' He turned to June. 'Can you see my mother saying go off to the woods for three weeks?'

'No,' June said, shaking her head.

'She was bamboozled.' He leaned over and looked at the letter on June's lap. 'It doesn't say anything about the parents not coming up to visit. At every other camp I was ever on, every weekend the parents were there. Not only did they deliver you and pick you up, they came up for events.' He shook his head. 'Three weeks!'

Doug Griset, long-standing judge at the Schenectady County Family Court, put on his robes each day and entered the theatre that was the courtroom, the drama of family dysfunction played out daily in front of his bench. So good was he at the role that he was cast as a judge in *The Place Beyond the Pines*, a gritty movie about a carnival drifter who washes up in Schenectady, which starred Ryan Gosling.

Doug was used to playing roles. But what he didn't know until he got my email was that it had begun in childhood, when he was cast in a role not of his choosing and without his consent. He was curious enough that when I suggested I call in to visit on my way back from a conference in Montreal in 2012, he suggested we could take a road trip to see if we could find the camp, and I jumped at the chance. I hoped that on the journey more of Doug's memories would surface, and I could make a connection between the scant detail provided by Sherif and his colleagues and Doug's personal recollections. I hoped it would prompt Doug to recover more of his camp experience.

I arrived in Albany, about half an hour from Schenectady, by bus. At the Greyhound terminal, Doug was easy to spot. In an artfully rumpled linen jacket, he looked coolly crisp among the people sprawled on seats, fanning themselves in the heat.

In Doug's car he spread out a map of upstate New York on the dashboard. Doug had been sceptical when I told him the camp I was looking for was in Middle Grove because it was only an hour away from Schenectady. 'I'm amazed because in my mind it was a thousand million miles away. It's like my dog when we go to our lake house, which is two hundred miles away. He sleeps most of the way because in his mind it's five thousand miles away and takes a hundred hours to get there.' He smoothed out the map. 'Now, I'm going to have to be a little careful here …'

The conversation in the car went mostly this way: staccato reflections and reminiscences interrupted by street directions.

'Here's where we're going.' When we got closer to the area, he pulled out some pages printed from Google Maps and smoothed them flat on the dashboard for me to see. He had his finger on a spot not far from Saratoga Springs. The indicator ticked as I looked at the map, not sure, now that I was here, what exactly I hoped we'd find. We had jokingly called this a road trip and, anticipating only Doug's memory as a guide, I had expected to find myself deep in the Adirondacks, probably lost, the outing fruitless yet enjoyable, an opportunity to talk and hear his story more than anything else.

We took three wrong turns on our way out of Saratoga. At the edge of town, Doug pulled over and went into a garage to ask directions. I wondered at the fact that we were geographically so close, but Middle Grove seemed hard to find.

When we found Middle Grove — not much more than a crossroad and a general store — we meandered. We followed minor roads up green hills and down, passed horse studs, dairy farms, and mansions set back from farm picket fences with sweeping gravel drives. We scrutinised signs.

We stopped at a gas station, but it was closed. Then we stopped at a store selling candy, a low, gloomy place where a woman emerged from the shadows and shook her head when Doug started his story. To an outsider, it sounded strange. He began by telling people how I'd come all the way from Australia and that we were looking for a camp that was run during the 1950s. Even to someone logical and trusting, it sounded like a cockamamie story. A camp, a group of boys, an experiment. It surprised me that people wanted to help us at all.

We passed Boy Haven, a summer camp for boys, but it wasn't the place. We got excited when we saw a sign for Camp Wood, thinking this might be it, yet were disappointed when we couldn't find an entrance. Then, after seeing three more signs for the camp of the same name, we realised they were referring to kindling for campfires. Finally, on a back road which I later found out was called Lake Desolation Road, we passed a single building: the Middle Grove post office.

Inside, Doug explained our story to the postmistress and I smiled and nodded in the background, convinced after two hours of driving around that this was a wild-goose chase. When Doug mentioned the camp and the experiment on the boys, she asked, 'Were you one of the boys?'

'Yes,' he said, but perhaps he saw something in her face because he hurried on. 'I'm all right. Nothing bad happened.'

Two customers at a bench in the corner behind us, wrapping up parcels, looked up at the exchange. 'Hang on,' one of the women said. 'I'll call my dad. He's ninety-five and has lived here all his life. He might know.'

Five minutes later, we were following her car out of the car park. Doug said, 'Okay, we may have hit a jackpot. We'll see. I'm still cautious.' But I could sense his excitement.

Two miles from the post office we turned up a dirt road, then off an even smaller track that curved in and out of tall stands of forest. Was this looking at all familiar to Doug, I wondered. But I could tell he wasn't thinking that far ahead. This was an adventure, and the thing with adventures is that you never know which way they will turn out.

At the top of a hill she slowed and pulled off to the side of the road, and Doug did the same. 'We would never have found this spot ourselves,' Doug said. We'd stopped at a fence facing a wooded area. A truck was parked beside a wooden shack, and a couple of cars were beside it.

'I see a guy with a dog up ahead, so we're going to get out, but we're gonna be real careful, right?'

A menacing dog ran towards the gate, and I noticed that a man in a flannel shirt stood on the porch, staring at us.

After a minute the man approached the fence cautiously, but made no move to call off the dog. The woman from the post office approached with Doug and told the man her father recalled a summer camp on this land in the 1950s. I waited by the car. Doug's comment about being careful reminded me how many people owned guns and might be suspicious of strangers. Then Doug started talking, telling our story. I moved towards the fence, and the dog reared up again, barking protectively.

The screen door slapped, and a woman in her twenties, in shorts and a t-shirt, with long dark hair, stepped off the porch. Her white legs flashed against the greenery as she made her way to the fence to listen.

The sun was dappled by the thick trees. A mosquito buzzed by my ear. After a moment, the man opened the gate and Doug waited for me to catch up before we stepped through.

I felt suddenly nervous. Doug's comment, reinforcing that we were strangers here, was right. We *were* strangers in this muggy green patch of forest at the top of the hill. The man didn't seem to know anything about a camp and didn't have much to say. The girl was also silent, watching from the fence line.

Doug had pulled out his business card, and was telling the man and the girl who he was. He was used to putting people at ease. Meanwhile I looked around. The narrow piece of land was small, the wooden house close to the road. It seemed impossible this was the isolated campsite Sherif had described.

The whole enterprise, our cautious excitement in the car, began to seem ridiculous. I was wasting everyone's time: Doug's, these locals', the women in the post office, the ninety-five-year-old father who seemed to remember the camp. But Doug had stopped talking and was staring at the house. 'That chimney,' he said to the man, 'is it inside the house, too?' Then almost straightaway, he continued, 'I remember a chimney, a stone chimney, with a big lintel where they displayed the prizes us boys would win.'

The girl invited us in and we stepped into a living room that opened off the porch. Doug pointed at the wall. 'That's it. They kept a barometer thing up there, with the scores of both teams in red and the prizes, these big jackknives that they paraded around. Every day we looked at how we were doing and we looked at those prizes. I wanted that knife so badly.' He shook his head. 'I can't believe how badly I wanted that knife.'

The girl asked Doug who ran the camp and he told her Muzafer Sherif, with the support of Yale and the Rockefeller Foundation. She wrote down Sherif's name on the back of his business card, Doug spelling it out for her.

I asked the man if he'd mind if I took some photos. 'Go ahead,' he said gruffly, and I stepped outside to the porch. From this vantage I could see the yard went back quite a way, ending at the edge of a forest of dark trees.

I snapped a few quick photos of the house and the land. Then I stood on the top of the rise and looked down the gully, where the land dropped away. A small wooden hut was half hidden in the trees. Doug came over and looked, too.

'It's an old outhouse,' the man said.

Doug nodded. 'A double outhouse, huh,' he said, and he

sounded as if he wasn't guessing. As if he knew the building too.

Below the outhouse, a wire fence was just visible in a tangle of bushes at the bottom of the hill, but if I half closed my eyes I could ignore the fence and see the site as part of a much larger property, accessible from a bumpy, unmade track.

Inside the house a kettle whistled and the girl abruptly switched it off. The man was hovering. We were holding them up. Neither he nor the girl could tell us anything about the history of their home. Doug had fallen silent, as if he'd come to the same realisation as me. There was no dramatic revelation to be had here. This place was not going to give up its secrets.

'She's in there googling Muzafer Sherif right now,' Doug said with a laugh as we pulled away. It had all happened so quickly that neither of us could quite believe how easy it had been, and as we drove we went over and over the sequence of events. What were the chances of being in the post office at the same time as the customer whose father remembered the site of the old camp? Wasn't it amazing that she had gotten in her car and taken us there? Doug was bowled over by how helpful people had been. I was still processing what we'd seen, but was increasingly sure we had found the right place. Despite the changes to the house, how incredible that Doug recognised the stone chimney as the mess hall where the boys had gathered at mealtimes. But Doug, like the judge he was, remained cautious. Yes, he recognised the stone chimney. Yes, he remembered the outhouse. Yes, it looked like it could have been the place. 'I know you want me to say it was,' he said, 'but I can't do that.'

Yet this trip had set something in motion in both of us, and raised more questions for Doug than I had answers for. What had his parents been told exactly, he wondered. Who were the men behind this experiment, and what had they been trying to prove? How much had parents such as his known about what they were sending their sons away to do? How exactly was he experimented on? Did he carry within him some hidden, unconscious legacy from the experiment, or was it just a couple of dimly remembered weeks of holiday?

I felt guilty, grubby, for stirring up these questions and having inadvertently started a process whereby Doug began to re-examine his life in view of this new information.

By the time we got back to Schenectady later that afternoon and he recounted the day's events to June, I knew that our discovery of the site of Sherif's mysterious footnote, instead of putting something to rest, was just the beginning. And I felt I owed it to Doug to find some answers.

# Part Two

# 4
# The Watchers

The cabins and the mess hall were surrounded by long grass when Marvin Sussman first visited the campsite in early June 1953. The Camp Fire Girls, the Girl Scout–like group that owned the property, hadn't used the place for a while and it felt abandoned, the small cabins hunched in the shadow of the woods. But Sussman was getting desperate: so many local campsites had already been booked out that he didn't have much choice.

Later in life, Sussman would usually wear tinted aviator glasses and a Stetson hat, but in 1953 he was clean-cut, with heavy black glasses and a plump, boyish face. He was one of five men Sherif had employed as part of his research team that summer, and it was his job to get things ready before the others arrived, recruiting staff, choosing subjects, and making sure the campsite was up and running. He'd worked with Sherif before, in the 1949 study — the one that took place in Litchfield Hills in Connecticut, where Sherif engineered conflict between two groups and had difficulty eradicating it, the one that gave Sherif the idea to look at ways to engineer peace between hostile groups

— and found him an inspiring figure. In that study, Sussman had been a participant observer: a person who pretended to be a camp counsellor but whose real job was to surreptitiously watch and make notes on what the boys said and did. But this was a much bigger undertaking, with more responsibility, more funding, more staff, and more pressure. And even though Sussman had sought out the job with Sherif, he was beginning to feel the full weight of it.

He wiggled the key in the padlock and the hinge groaned when he pushed the door open. Inside, the mess hall would have smelt damp and musty, and I could imagine Sussman propping the door open with a block of wood and taking a moment to fill his pipe and light it, clamping it in his teeth as he stepped inside and looked around. Since he last worked for Sherif, he'd completed his PhD and now had his first academic job, at Union College in Schenectady, but he felt unappreciated there and was desperate to get out of the place. This study was his chance.

Sherif took Sussman on because he was a hard worker who could organise the practical details. When they met, Sussman was a Yale graduate student who worked eighteen- to twenty-hour days supporting a young family. As well as studying full-time, he worked part-time as a research assistant, worked nights at a diner, and took in extra work helping out in his father's business in watch and clock repair. When Sherif offered him the job of research collaborator for his upcoming summer research, and co-authorship on a book about it, Sussman jumped at the chance. Sherif, a leading figure in social psychology, with influential connections, was giving him a boost up the academic ladder, a way out of a small college to a bigger city, a better job.

Sussman started working for Sherif in May 1953, and with the experiment planned to begin at the end of June, he threw himself into the detail of arrangements. His job was to be on the ground, organising the logistics under the approval of Sherif, who was down in Oklahoma. It might have sounded straightforward enough, but Sherif could be an infuriating boss, prevaricating on major decisions and micromanaging small ones. While he promptly fired the camp nurse Sussman had hired because she was too pretty and likely to be a distraction to the men, Sherif delayed giving the go-ahead on booking a campsite. With just four weeks before the experiment was due to start, they had no campsite booked, and without a location for the experiment or a confirmed date, Sussman was unable to recruit staff or select subjects.

By the time Sherif agreed on a site, most of the families Sussman had contacted had already signed their sons up for summer camp. They moved the camp date forward so that Sussman had more time to complete arrangements. His letters to Sherif in the weeks preceding the camp were increasingly plaintive; he was worried that so much of his time was being squandered on mundane organisational tasks rather than on developing the theoretical framework for their book. Sherif's wife, Carolyn, urged her husband to put himself in Sussman's shoes to understand his frustration. 'You must be able to imagine how dreadful it must be to be stuck in a little one-horse college with 3 children where no stimulation or accomplishment is possible and with just enough to make out on …' But Sherif was unsympathetic. This kind of hard work was the price of groundbreaking research, he wrote to Sussman, calling the task 'Herculean' but reassuring him that he had the sort of spirit needed to 'succeed in this "frontier" attempt'.

For Sussman, perhaps this demanding and often frustrating supervision was the price of working with a genius. He persisted, and by mid-July, after contacting fifty-three local ministers and thirty-five school principals, and cold-calling two hundred and twelve parents, Sussman had twenty-four boys signed up.

The six members of the research team who converged on the campsite on 18 July 1953 hit the ground running. With just five days before the boys arrived, the men were up early each morning and finished late at night. Even if there'd been time for long walks in the woods or swimming in the lake, Sherif would never have allowed it. Renowned for a ferocious work ethic, he expected nothing less from his staff.

By day, amid sounds of sawing and hammering, as workmen connected electricity, gas, and phone; installed plumbing for running water; and drained and repaired the dam so it could be used for swimming, Herb Kelman sat inside the stuffy cabin the staff had commandeered as an office, typing furiously. Sherif had invited the twenty-six-year-old he'd met at Yale to be research consultant, the scientific conscience of the study, ensuring that the men were rigorous and objective in their approach. I had never heard of someone being invited into a research to play this kind of role before, and it seemed to be an innovation of Sherif's. Kelman was already making his name as pacifist who urged his fellow psychologists to apply their expertise and the discipline of science to advance the psychology of peace, not war. His role in the experiment was to develop scientific standards for the experiment and, as a roving observer, ensure the men kept to them.

But Sherif and Sussman had to do more than demonstrate their objectivity in their observations of the boys to prove the experiment was a success. They needed measuring devices to gauge the amount of trust and loyalty between the boys. This was to be the PhD research project for OJ Harvey and Jack White, two of Sherif's graduate students from Oklahoma. They spent most of the five days stripped to the waist, hammering and sawing in the shade of the mess hall, constructing a ballgame and target board that would act as a friendship-measuring device.

The sixth man was Jim Carper, who Herb Kelman had recommended when Sherif was looking for another observer. Carper had plenty of practical skills developed from working in conscientious objector camps during the war, and during those five days he helped out the others, mowing grass and chopping wood in preparation for the arrival of the boys.

It seems an idyllic portrait, six men working alongside one another to prepare the scene for one of the most complex and daring field experiments of the era. Yet, much as they would with the two groups of boys, tensions were already brewing between the experimental team. Sherif's letters in the months leading up to this time were full of dread and foreboding, and he seemed unable to trust Sussman to do a good job. He was convinced that this research was the most challenging and important project he had ever embarked on, and securing a huge grant for it — $38,000 from the Rockefeller Foundation — seemed another testament to its merit. The grant was one of eight the Foundation funded, and surpassed the American Museum of Natural History's study of community life on Manus Island, led by Margaret Mead. Yet instead of filling him with confidence, the grant seemed to have done the opposite. In March 1953, he had written to his mentor, Carl Hovland, that the experiment 'is such a huge job that the burden of responsibility for doing justice to it is weighing heavily on me'. Perhaps Sussman, as the leading man on the ground, was the recipient of Sherif's anxiety.

In contrast to his rather gloomy letters to Sussman, there's a kind of longing in Sherif's letters to his graduate student and research assistant OJ Harvey as they negotiate how soon Harvey will be able to join the team, with Sherif urging him to hurry and get there as soon as he can. I imagine when Harvey's blue 1949 Plymouth rumbled into camp after the long drive from Oklahoma that Sherif rushed out to greet him and Jack White, before immediately setting them to work.

On Kelman's advice, Sherif had divided the six men into the research team and the operational team, kept physically separate and housed in different cabins to guard against bias. The research team, with Sherif as director, included Kelman, Sussman, and OJ Harvey. They had a small, two-storey cabin to themselves, plus a tent in the birch grove, particularly for Sherif's use. The two participant observers, Jim Carper and Jack White, who would spend most of their time with the boys, shared a cabin of their own.

For those five days, from 18 to 22 July, before the camp began, the six men worked on their roles. Two would pretend to be caretakers, two would be camp counsellors, two would be camp managers. They memorised the main hypotheses for each stage: that groups would develop their own identity and sense of belonging during the first stage, that each group would develop negative attitudes towards their rivals during the competition phase, and that group friction and boundaries would dissolve when the groups faced a problem that required them to cooperate to find a solution — a forest fire that threatened the camp — in the final stage. The men also practised and tested their observational skills and perfected their disguises, taking turns in snapping group photos with the new cameras Sussman had bought to record the experiment. The suave-looking Sherif of earlier photos I'd seen had been replaced by a rather portly middle-aged man wearing a rumpled grey janitor's uniform. His wife, Carolyn, a budding social psychologist herself, who did all sorts of behind-the-scenes work for Sherif, including co-writing books and journal articles and soothing his professional anxieties, likely chose the outfit for him.

It's hard not to read something into the photos the men took of themselves in the days before the experiment began. I noticed how the only photo of Sherif in which he looks relaxed is the one where he stands between his two graduate students from Oklahoma. His hands in his back pockets, smiling around a cigarette hanging from his lip, Sherif is half turned to a handsome young Billy Jack White, or Jack, as he was known, who wears a tight white t-shirt and an army cap pulled down to shade his face. With his olive skin and his sleek black hair, Jack White reminded me of one of the Italian boys from *West Side Story*. He looks as though he is telling Sherif a joke or a funny story, a cigarette pinched between his fingers and half raised to his mouth. On the edge of the photo, with his back to them but facing the camera, OJ Harvey smiles along as if he is anticipating the punchline, his hands on his hips and his eyes on the horizon. OJ's brown hair, Brylcreemed in a series of sculpted waves, makes him look dependable, someone you can rely on when things get rocky.

The men that stand beside him in the photos were Sherif's bulwark against failure, each of them chosen to play a particular role, all of them at the start of their careers, unlike forty-seven-year-old Sherif. In one photo, Marvin Sussman, with his fleshy face and round glasses, a baseball cap tipped back on his head and swinging a box brownie from one hand, looks like a dad about to head off to a baseball game. On the other side of Sherif stands Herbert Kelman, wearing a plaid shirt and a leather cap, the brim pulled low over his heavy black glasses. With the stubble of a black beard coming in, he looks swarthy and slightly disreputable. The plan was that Sherif and Kelman, posing as caretakers and shuffling about the campsite picking up rubbish, raking leaves, or chopping wood, would be able to move freely from group to group, boy to boy, making observations and 'asking naïve questions', supposedly without drawing attention to themselves. But Sherif, if not Kelman, had been seriously miscast. He had grown up with servants and had never been a handyman, had never used a screwdriver or fixed a squeaky door.

I had already encountered a Harvard professor called Herbert Kelman whose work on ethics I had come across in my research on Stanley Milgram. But as hard as I looked online, in his lengthy curriculum vitae and in interviews about his life and work, Kelman made no mention of working with Muzafer Sherif. I was intrigued.

'That's me,' Herb Kelman said when I showed him the photo of the group of men, pointing at the rather shady-looking character with the leather cap and the dark stubble. 'I decided not to shave for the duration of the camp.' With his white hair swept back from a bald pate, Kelman didn't look anything like the man in the 1953 photo, except that he wore the same kind of heavy-framed black glasses. 'I was known as Mr Herbee, a handyman,' he said with a chuckle.

Two years before that photo was taken, in 1951, Kelman, a graduate student at Yale, had been 'very excited' to hear that Muzafer Sherif would be visiting the university. He'd been a big admirer of Sherif's work. Despite the age difference — Sherif was almost twice Kelman's age — the two hit it off. Sherif had just finished his first 1949 study, where he had brought two groups to conflict, and, when they met, Sherif was developing a theory for the next stage: how to bring about peace. 'I liked his work very much — he was beginning to develop his theory of conflict and moving from hostility to harmony. His general approach was to look not at individuals but at the relationships between groups of people and the social context in which violence and hostility occurs.'

Kelman's interest in peace research was personal. As an eleven-year-old in Vienna, he had witnessed the arrival of Nazi troops after the Anschluss and the horror of Kristallnacht before he and his family fled to America. Sherif had asked Kelman for feedback on a chapter he had been writing, and Kelman noticed how insistent Sherif was that conflict and war was not the result of human nature, of a person's authoritarian upbringing, or of frustration in early childhood. 'He spent an awful lot of time in his writing tearing down other approaches, particularly psychoanalytic explanations of groups and group behaviour. And I tried to tell him, "Why do you do that? It isn't really necessary.

You don't have to tear other theories down in order to come in with your own. You could say yes, here are other theories, they have their limitations, this approach is an attempt to deal with those limitations." So I stress this because it became an issue in the later work. He had a fatal flaw. He could never admit when he was wrong.' But Sherif admired Kelman's frankness and invited him along as a kind of safeguard against scientific bias.

As Kelman would soon learn, for Sherif, inviting advice would prove much easier than acting on it.

Nine boys stood in a ragged line in a clearing, holding their bows and arrows. An impatient queue of six other boys watched from the sidelines, waiting for their turn. Peter Blake held his bow expertly, and was trying to get the others' attention, calling out loudly to watch how he did it. Almost thirteen, Peter was small for his age, but he had the easy confidence of a boy who was used to being in charge. He stood with one hand on his jutted hip, frowning impatiently. Some boys were shooting already, others were still trying to get their arrows correctly in the bow. Peter called that they should do it all at the same time.

It was the first day of the camp, the first time Doug, a slight, sandy-haired eleven-year-old, had held a bow and arrow. He told me how he loved archery: and what boy wouldn't? It was a thrill to feel the tautness of the bowstring when he pulled the arrow back, the tingling anticipation of letting the arrow fly, even when it fell short and hit the ground instead of the target. I can picture him waiting patiently with the others on the sidelines until it was his turn again. He loved the danger of it. When he went back home after the camp, he pestered his parents to be able to practise archery in the backyard. But their answer was an emphatic no.

Peter lifted his bow and inserted an arrow. 'Ready,' he called and squinted along the arrow, aiming it at the target ten metres away. 'Aim!'

Just then, a smaller boy darted out in front of the line to retrieve his arrow from where it had fallen short of the target and lay in the grass. Another boy let his arrow go and it whizzed towards the target. Peter stamped his foot. 'Hey!' he yelled angrily. He spotted Jim Carper, a camp counsellor, lounging against a tree by the edge of the woods, smoking a cigarette. 'Jim!' Peter appealed to Carper. 'Can you call the commands?'

Twenty-eight-year-old Carper, with the quiff of dark hair that flopped onto his forehead, James–Dean style, called back, 'You're doing fine!' and gave Peter a lazy thumbs-up.

Peter made an exasperated noise and turned away.

Beneath his casual pose, Carper was watching intently, memorising the scene so he could write it all down later. He described this exchange in his daily observation notes as proof for his new boss Muzafer Sherif that he was fulfilling his role as disinterested observer. Carper was the only one of the research team who hadn't met Sherif before the experiment, and I imagine that at the same time as he was trying to get the measure of the man, he was eager to prove he would do a good job. I don't know if it was money or idealism that brought Carper to this camp — or if it was the chance to work with a renowned social psychologist — but he had had enough of the rat psychology he had been studying at John Hopkins. After the sour smell of the animal laboratory, the warm Adirondack air, cut with the smell of mown grass, would have been invigorating.

Today, the boys had free run of the campsite, and it was the job of the six men disguised as camp staff to watch them mingling and playing and make note of who was making friends with whom. But reading Carper's description of the chaos and exuberance of the archery range on the first day at Camp Talualac, it's impossible not to feel worried for the boys' safety and sympathy for frustrated Peter Blake. In his attention to observing detail, Carper seemed oblivious to danger.

The boys' over-exuberance in the archery session was a release of pent-up energy. It had been raining most of the day and they'd been stuck indoors. The rain had started just before the bus had pulled off the muddy dirt road and into the camp, and there'd been a scramble to get their bags off the bus and into the mess hall without getting drenched.

Doug waited his turn, watching the boys around him fooling around, shouting encouragement, and boasting about how they'd hit the bullseye. He would have felt puny in comparison. He was a 'dweeby' kid, he told me, who looked younger than eleven — a fact his father, who had fought during the war and now ran a pharmacy in Schenectady, was always trying to help him compensate for. He made Doug wear an oversized coat in the schoolyard to help him look bigger and gave him boxing lessons. But Doug had already intuited that physical size wasn't what got you respect in the schoolyard. It was personality. So he didn't boast as he waited his turn, didn't brag like some of the others about how close he got to the bullseye.

It had seemed like a long trip to eleven-year-old Walt Burkhard. Then again, a trip to Schenectady, to go to the library with his mother, from their small village of Alplaus seemed like a big deal too. Schenectady felt like a 'metropolis' compared to Alplaus, which had just a post office, a garage, a general store, and a tiny four-roomed schoolhouse. But Walt was shocked by how isolated this camp was. There was no town nearby: they were right out in the woods with just a big dark mess hall and five drab cabins, surrounded by thick woods overgrown with vines that draped from the trees like cobwebs. It was the first time he'd been away with a group of people he didn't know. And in a small place like Alplaus, Walt was used to knowing everyone.

After a welcome by the camp director, an introduction to the staff, and an inspection by the nurse, the boys were encouraged to claim one of the twenty-four beds that had been set up at one end of the mess hall and unpack.

The wet weather might have ruined Sherif's plans for the day's activities, but it was a stroke of luck for those boys who felt far from home. With twenty-four children and ten adults inside and the rain falling steadily outside, the large room was warm, crowded, and cosy.

Walt and Doug quickly made new friends. Doug found a buddy in John Wilkinson, who on the bus had impressed them all by knowing at a glance the make and model of every car they passed. That wet afternoon, John and Doug lay on their bunks, reading comic books together. Walt overcame his shyness and joined in a game of cops and robbers after a serious-looking boy in glasses called Irving asked him to play, and a group of them ran around the mess hall, using sticks as guns, while around them other boys wrote letters home, or played Parcheesi or Crazy Eights. Meanwhile, Harold McDonough, a boy with an interest in all things scientific, sat alone, and I imagine him reading a copy of *Popular Mechanics*, with its offers of correspondence courses in industrial electricity, build-your-own power mower kits and crystal radio sets. He kept himself to himself. He was the kind of boy with an interest in how things worked and expected adults to be able to explain things and answer his questions. Except that he noticed that the adults here had trouble explaining things. When he asked Jim Carper that first morning what the microphones in the rafters in the mess hall were for, Carper stammered and said they were there for when Mr Ness made camp announcements. But they were microphones, Harold pointed out, not loudspeakers. Jim just looked uncomfortable and hurried away.

In the kitchen, divided from the main room by a partition, the cook and his helper fried chicken croquettes, hash browns, and buttered carrots, and boiled green beans for lunch. The air was steamy and fugged with food smells, the heat of warm bodies, and the men's cigarette smoke.

After lunch, one of the counsellors brought out his mandolin, and two boys — Eric on the accordion and Irving on the ukulele — joined in to play 'On Top of Old Smokey'. Irving's shoulders hunched, and his glasses slipped closer and closer to the end of his nose as he stared intently at his hands as he played, so determined to get it right that he forgot to sing. Eric played enthusiastically, not noticing or not caring that he was playing out of tune. A boy named Laurence joined in, substituting the lyrics to 'On Top of Spaghetti' and waving his arms to get the others to collaborate.

By mid afternoon, the rain had cleared and the boys headed out to the archery range. A hot breeze dried what was left of the muddy puddles. Irving was having trouble with archery. The humid air fogged up his glasses, so it was hard for him to see the target. Despite never being able to hit the target, and despite the urging of some of the boys for him to give up his turn, Irving persisted, lining up again and again but never getting any closer to the bullseye. At supper, he saved Walt a seat beside him. After supper, John and Doug sat on the jetty and John showed off his knowledge of fishing to his new friend, demonstrating to Doug how to bait a hook before casting the line out into the stream. They fished together until it got dark.

It was hard for the boys to sleep that night, even though it had been a full day. There was a lot of giggling and excited talk after lights out. Although some were quiet, trying not to think about home, others bragged about their exploits in archery, and a few told knock-knock jokes or shouted mock insults in the dark. But soon one of the camp counsellors came in and told them to settle down. Gradually the talk died, as one by one the children fell asleep. Jim Carper, listening at the door, noted what time it was when the room went quiet and the last boy had gone to sleep.

The next morning the sky was clear, the puddles had almost disappeared, and the only evidence of rain was the mosquitoes swarming in the shade of the trees at the edge of the clearing. After a breakfast of ham and scrambled eggs, Marvin Sussman rounded everyone up outside the mess hall to take some group photos.

Sussman was the only adult the boys had met before. He had visited their homes to talk to their parents about the camp. I imagine they lined up eagerly and followed Sussman's directions: taller boys in the back row, the front row seated cross-legged on the grass. They were a neat, well dressed bunch. Most of them looked as though they had had their hair cut for camp, and wore bright striped t-shirts, or neat checked shirts tucked into jeans. Many looked relaxed, expectant. Some looked shy. You wouldn't know, looking at the photo, that just the day before they were strangers.

But like everything at this camp, the photos had another purpose. Once they were processed, Muzafer Sherif pinned the group portraits to the wall of their small office so the men could more easily memorise the faces and names of their subjects.

After the photos, the boys raced towards the archery range, while those who missed out played games of Horseshoes while they waited for their turn. Even though they had only been at camp a short time, the boys were already forming strong connections. Sherif predicted that the boys would make friends during this first stage, and the observers' job on the first day was to record 'emerging friendships'. But it seemed from the documentary evidence that there was nothing fledgling about the camaraderie that was developing between the boys. I wondered if they unconsciously understood that there was something different about this summer camp, if it was a reaction to adults who did so little to organise them and yet seemed ever present and strangely watchful that the boys instinctively bonded.

For the boys, the second day of camp seemed an extension of the first. At the archery range, new buddies Peter and Laurence had established a system and set up rules, and the rest of the boys seemed happy to follow their lead. They made a good team: Peter, with his bossy and rather serious manner, kept order, and stocky Laurence, the comic, with his gift for funny faces and knock-

knock jokes, entertained the others while they waited their turn. The boys now formed a straight line and fired in unison and only on command.

That afternoon, the campers went spontaneously from one activity to the other, unfettered by adult intervention. The boys believed the camp was run by Harry Ness, an emaciated man with a mournful face who wore small shorts that exposed startlingly long pale legs. Ness carried a clipboard and wore a whistle around his neck, and it was he who announced the day's activities and gave instructions to the two junior counsellors, students at Union College, Ken Pirro and Rupe Huse, who spent most of the day with the boys, supervising activities, supplying equipment, and generally making sure things ran smoothly. Then there were Jim Carper and Jack White, introduced as the senior counsellors, who also spent most of their day around the boys. But in the shadow camp, Ness and his team were puppets whose every announcement, every move, was dictated and choreographed by Muzafer Sherif, whose plans Sussman passed on to Ness and the others each morning before breakfast.

Sherif's strategy was for the boys to pass through four stages. The first stage was mingling and making friends, or 'spontaneous group formation', and it would last just a day or two. In the second stage, 'intragroup relations', the boys would be separated into two groups and each group given a chance to develop its own identity. In stage three, 'intergroup relations', the groups would take part in a series of competitions that would cause hostility and conflict between them. Stage four was the 'integration phase', where the two groups would come together in the face of a larger, shared problem. The hostility and competition between the groups would vanish, and they would regroup as a harmonious whole. That was the plan.

At every stage in these first days, watchers such as Carper were on hand, observing, writing notes, and recording on film and audiotape when they could. But how do you transform the jostling, noisy, and chaotic interactions between groups of children into a set of scientific observations and data that others will understand? How do you show that the hunches you have are borne out by your observations?

In the five days before the experiment started, Herb Kelman hammered out three drafts of detailed instructions on his portable typewriter, and they are an eye-glazing read. As well as a staff policy that included a rule of no drinking, Kelman wrote five densely typed pages of advice for the participant observers in how to maintain a hands-off approach. 'Since the observers are familiar with the hypotheses, they may tend to expect certain kinds of behavior … and may be selectively perceptive … They should try in every way possible to counteract this tendency …' The instructions were full of warnings about what not to do. Page one told them, 'Nobody is to be a leader to the boys.' Staff were to remain low-key and were not to distinguish themselves in any way that might detract from the boys' relationships with one another: 'We do not want boys to develop attachments to certain staff members …' They were not to influence the behaviour of the boys in any way, and in particular not to usurp the role of the groups' natural leaders. To this end, staff were not to make suggestions, question the boys, display special talents, form attachments, or wear insignia. Instead they were to be unobtrusive observers, taking photos and movie film covertly, memorising their observations so that later, when they could get away from the group, they could write notes without arousing suspicion.

In addition, the counsellors were not try to 'influence' campers, were not to take initiative 'in introducing activities', and were not to 'counsel campers'. They were also not to initiate anything without the direction from research staff. 'It is better to do nothing than to begin a course of action which may prove deleterious to the operations of the study,' the booklet makes clear. Kelman also warned that the research team would be watching to make sure they complied. They were reminded, too, that the camp could well take an unexpected turn, but that no matter how strange, 'Be assured … that it is done with the best judgement regarding the success of the project and the participants.'

But reading these instructions, I was struck by what they didn't say. How, in practical terms, could the counsellors implement these directives given they were the ones supervising activities and spending all their time with the boys? It seemed an impossible ask.

On the first day of camp, the researchers' goal had been simply to record friendships, such as Peter and Laurence's, and Walt and Irving's. On the second day, they were preoccupied with assessing the boys' skills in a range of activities so that later, when the boys were split into groups, the two sides would be equally matched. But the men seemed to have underestimated or perhaps didn't notice just the strength of the bonds developing between the boys.

Walt had no idea that he and his friends were being studied — he didn't notice anyone observing them or taking notes. Initially the camp was just as he imagined a normal summer camp would be, although that afternoon none of the adults comforted a boy who had the 'sniffles' when it came to swimming, and no one encouraged a quiet boy called Tony who seemed 'reluctant to mix'.

As the second day wore on, it became clear to Sherif as he roamed the camp, dressed in a janitor's uniform, that assessing the twenty-four boys spread out across the woods and grounds was too difficult. I pictured Sherif prowling the campsite pretending to pick up litter, watching with irritation at what he would have viewed as aimless activity as the boys — some in groups of two or three, others on their own — drifted from one pastime to another: tinkering at an old piano outside the mess hall, shooting arrows, playing badminton, and tossing horseshoes. He decided they had to change plans. The only way to successfully differentiate and compare the boys was by getting them in the same place at the same time to play a game.

After lunch, when Ness suggested a soccer game, the boys 'rejected' this. Fifteen minutes later, the junior counsellors 'forcefully' rounded the reluctant boys up for a game of volleyball, although four boys still refused to play. With Ness as umpire, the six observers watched from different vantage points on the sidelines, ready to rate each boy in terms of size, sporting ability, and leadership potential in games. But it was sweltering in the sun; the boys were tired from the morning's activities and 'unenthusiastic' about playing. Next, Ness announced a game of softball and took the boys to the baseball field, but when they got there the boys again refused. Finally, most of the boys agreed to a game of prisoner's base, but it soon degenerated into 'general apathy and chaos'. Carper wrote that Ness' instructions for how to play were confused, and once play did start, 'many of the captured prisoners accepted their incarceration as an opportunity to sit down and do nothing', and were not interested in being rescued. Teammates such as John and Doug sat in the shade and read comics while Harold 'went hiking in the woods, alone'. Eventually the men gave up and the boys wandered off to swim, play badminton, and practise with the bows and arrows. I can imagine Sherif's irritation that such a seemingly straightforward exercise in observation had proved so difficult, and his unhappiness with the men's inability to inspire the boys to play. But the afternoon had highlighted a weakness in the researchers' view of their subjects. The boys were not passive pawns, and their cooperation was crucial to the experiment's success.

The second day of camp ended happily for the children. Around a campfire, they took turns in telling ghost stories and, later, joined in a sing-a-long. The whole group sang 'Hail, Hail, The Gang's All Here', but Herb Kelman later wrote in his notes that they all 'balked' at singing the word 'hell' in the line 'What the hell do we care' because it was a curse word.

By eight o'clock, the campers were in bed and settled. If any of them had stayed awake after the others had fallen asleep, if they'd listened hard, they might have heard above the steady click of crickets the *tap-tap-tap* of typewriters coming from a cabin on the other side of the birch grove as the two watchers typed out their observations of the day. Side by side, Jim Carper and Jack White typed without speaking, forbidden from discussing their view of the day's events with each other — their fingers flying faster as the time neared 10.30 pm, when they were scheduled to present their observations to Muzafer Sherif.

Sherif had specified this deadline, annotating Kelman's instructions with his pencilled scrawl. It was a pattern I recognised from his letters to Marvin Sussman: the adding of more detail to already exhaustive instructions. He dictated what time participant observers Carper and White must finish observing the boys each day, how and when and to whom they were to turn in their notes (Sherif), who would read their notes (Sherif and Kelman), and who they could and couldn't discuss their observations with. Pooling their men's typed observations with their own, the research team of Sherif, with Kelman, Sussman, and Harvey would formulate and finalise plans for the next day, rarely getting to bed before 2.00 am.

It felt like more than just a case of scientific scrupulousness at work here — something akin instead to an exaggerated fear of failure. Even acknowledging Sherif's anxieties about the need for the experiment to go well, his close supervision and desire to detail everything meticulously gave the impression of a man who didn't seem to trust the men working for him to get things right.

On the other hand, Sherif had an intellectual energy that swept others along. I can picture him striding up and down in front of his assembled team in the small room that served as an office, gesturing emphatically, his face glowing with heat and excitement, urging the men to remember the momentousness of this research, just as he had persuaded superiors at his university of the magnitude of his work. His staff were all young men at the beginning of their psychology careers, inspired and perhaps even a little overawed to be invited to work with a theoretician of his stature. If Sherif went on too long, if he repeated himself or sounded at times as if he was haranguing them, it was because they had not yet proved themselves. But did the men catch one another's eye if he went on like this, I wondered, or did they look away?

As Sherif read the men's typed notes, he marked points that needed following up with two dark pencil strokes in the margin, like exclamation marks without the points. In his notes on the second day, Carper naively included conjecture that went beyond observations of what he saw and heard among the boys. He noted the boys' 'apathy for organised sports', suggesting it might be due to poor selection procedures — a remark that likely would have annoyed Sussman, given how much effort it had taken him. Carper's comment would have bothered Sherif too: not just because it suggested scientific sloppiness but also because so much of the later conflict Sherif was planning revolved around competitive games. Perhaps Carper offered this observation to impress Sherif, not considering that it might light a fuse of antagonism.

Jack White's notes, on the other hand, could be describing a completely different day. There is no mention of the boys' reluctance or resistance towards group games. He stated which boys took charge and which boys were happy to take turns in archery, Horseshoes, and badminton. In White's notes, boys volunteered to carry sporting equipment and to be captains for each team. White observed which boys were the best players. In contrast to Carper's, White's notes indicated that things were proceeding according to plan. I wondered about this discrepancy and remembered the photo of Sherif, smiling and relaxed, flanked by Jack White and OJ Harvey, and how, unlike in the other photos, where Sherif stood apart from the others, in this one he looked connected to these two men, as if an invisible string ran around and between them, binding them together. Of all the men, these two knew Sherif best, and with both using the experiment as the basis for their PhDs, they had as much invested in the experiment as Sherif himself.

I pored over these two versions of the same afternoon. In the discrepancies between these accounts lay the problem of this kind of research — despite Sherif's best efforts to standardise the process, subjectivity could never be eliminated. But it seemed more than just a difference in point of view: the contrast between what Carper and White observed was stark. Jack White was thoroughly familiar with the details of Sherif's work; his own dissertation was based on Sherif's theory. From White's vantage point, things were running smoothly, whereas Carper's told another story. It seemed that already the men's observations were being shaped by their point of view. But whose version would Sherif believe? Perhaps it was as early as this, the second day of the experiment, that the rift among the experimental team began.

At the end of the second night, after he'd reviewed Carper's and White's notes with the others, Sherif took Sussman and crossed the campsite in the darkness to the small tent Sherif had reserved for his own use, tucked away at the edge of the woods. As they would every night from this point, they stayed up into the small hours, Sussman typing up the next day's events and Sherif refining and Sussman retyping. I can picture Sussman sitting and Sherif pacing in the tent, lit only by torch beam, discussing how they could make the boys interact. From the outside, the tent would have been an illuminated triangle, and even if one of the boys had seen it from the mess hall, they would likely not have realised anything was amiss.

Eventually the events were set. In my imagination, Sussman lit his pipe and began typing the script for the next morning's announcement as Sherif paced back and forth, smoking. It was a script that, for the boys, would change everything.

The next morning, Saturday 26 July, the boys gathered after breakfast, as they'd been told, outside the recreation hall. Their luggage — a pile of suitcases and duffel bags, blankets and pillows — were piled up on the grass. Harry Ness made a carefully scripted announcement, but even Sussman's final typed version Sherif had annotated with pencil, adding stage directions in parentheses:

> The boys can now move into their permanent quarters, which consist of two tents. In announcing this mention that … the new arrangement will be much superior: sleeping facilities won't be as congested, and also the rec hall will now be available for recreational facilities.
>
> Ness then announces: "The following boys will be together in one tent. As I read your names, come up and stand here (point to one side)." He reads the names of one "group". He then says: "The following boys will be in the other tent. As I read your names, come up and stand here (point to other side)."
>
> (Note) While the names are being read and the boys are lining up, movies and pictures should be taken. Observers should note the reactions of the boys.

I don't know what excuse, if any, the men used with the boys for taking photos this morning, although they are shot from a distance, as if the photographers might have been hiding themselves from view. There are a handful of pictures, as well as some film footage of this event. One photo is taken from the rec hall, looking down the slope to where Ness — tall as a bean pole, his pale legs glowing, wearing a white baseball cap — is speaking to the campers, who surround him in a half-circle.

The next photo is taken just a moment later, but from the front, facing Ness, after he has finished his announcement. The orderly half-circle is disrupted: some boys have stepped towards Ness, others have turned away. Most look dejected. Pairs that staff had noticed having fun together — including John and Doug, Peter and Laurence, Walt and Irving — have been separated and put in different groups.

Peter argued with Ness, telling him they should be allowed to stay with their friends and that he didn't want to be separated from Laurence. Laurence, usually quick with a joke, looked ready to cry. When Ness insisted, Peter looked disgusted and turned away. Three more boys asked Ness if they could swap groups, and Jack White noticed another two boys crying as they went to get their luggage. The boys were upset that Ness, who had been so permissive since they arrived, allowing them to pursue their interests and their friendships freely, was now breaking them up in an arbitrary way and ignoring their pleas to be in a tent with their friends. Ness, the one the boys were told made all the decisions about what happened at camp, was in fact the fall guy, the one Sherif hoped the boys would blame instead of the other staff for 'changes in policy … that might seem strange to the children'.

Sherif was jubilant at the boys' misery because his first hypothesis had been proven. *Note the displeasure of Ss* (meaning subjects), he later wrote on the back of one of the photos. The boys' sadness at being separated was a measure of their friendships. The stage was set for the next phase — what happens when friends are divided and end up on opposite sides in competing groups.

# 5
# Initiation

The tents were a quarter of a mile apart, on opposite sides of a stream. The boys reluctantly gathered up their bags and belongings, with help from the caretakers Sherif and Kelman, and took them to their respective tents. Carper's group was on the west side of the stream.

Harold, still carrying his *Popular Mechanics* magazine, wasn't surprised they had been split up. He had noticed the men watching as they played games the day before, and when he saw the junior counsellors conferring with Harry Ness that morning, he guessed what they had in mind. Harold told Jim Carper that he knew the men had been watching to find out who the boys' friends were and then separated them so they could make new friends. Carper insisted this wasn't true but he included the information in his daily notes. Yet Harold must have kept his suspicions to himself because Walt, who was in the same group, remembers the separation as a shock. 'We did everything together in those first couple of days. We got to know each other and developed a real sense of camaraderie.'

Looking back, Doug thinks their distress at being put in different groups says something about the situation the boys found themselves in. 'When you think about it, there were twenty-four boys and none of us knew each other. We didn't come from the same communities, neighbourhoods, or schools, so the fact that in just one or two days we became that friendly that we didn't want to be divided is telling. I'm wondering if the place was so darn spooky that we didn't want to be divided for that reason. Because this campsite was not your standard summer camp, with a pretty lodge and its little cabins or lean-tos. This was the woods, and you were a long way from your home, and I don't get the sense that the counsellors were trying to be our buddies and supportive, so all in all it had to be a fairly unpleasant experience for everybody. Maybe we just felt, "At least we have each other."'

Sherif wrote that after the boys had been moved to their new tents, 'the pain of separation was assuaged by allowing each group to go at once on a hike and camp out'. But his notes were a case of wishful thinking. Distracting the boys from being separated would not be as simple as that. Soon after the boys were taken to their new tents, the men realised that Mickey, a quiet, stocky boy, was missing. Harvey, Sussman, and Ness fanned out across the campsite, but Mickey was nowhere to be found. Harvey and Ness got in the truck and drove up the road, and Kelman's notes quote Harvey's description of the scene. They spotted the boy 'one mile from the camp in a fast run. He ran into the bushes and hid. When they got him he sobbed loudly all the way back. He said "The boys will call me a sissy" and "Please let me go home."' Kelman doesn't make clear whether they got the boy out by force or by persuasion, but it sounded as if they hadn't changed his mind. Harvey wrote in his notes that between sobs Mickey told him he 'didn't want to come to camp but his mother made him'.

When Mickey was returned to camp, Carper made a fuss of him, and in preparation for the day hike, 'I gave him the first aid kit to carry.' The boys were given haversacks, canteens, and mess kits, and some of the boys crowded round 'excitedly' to see them, Sherif wrote. Carper asked Mickey to help him assign the other boys their gear. 'This cheered him quite a bit,' Carper wrote. I was beginning to like this man. I imagined Mickey felt the same way.

The caretakers led each group on a day hike — the boys were told this was because they knew the area so well. Carper, junior counsellor Rupe, and Kelman, led their group of twelve boys — which included runaway Mickey, shy Walt, jovial Laurence, fisherman John, and *Popular Mechanics* fan Harold — north. By now Carper had memorised the boys' names and was free to watch how their relationships were developing.

His started off as a large group, but once the path started to climb, they strung out in a long line. At the head of the group, John, an experienced hiker, kept pace with Kelman, or 'Mr Herbee'. Behind him, in a loose group of four boys, Laurence, usually the joker, was subdued, flicking a stick at the long grass, saying little. Beside him, Mickey struggled to keep pace, sweating and puffing. The boys were clustered in groups of three of four, and Carper noticed how they avoided talking to Mickey. He had overheard the others complaining about Mickey, how he always wanted to be 'different', refusing to join in games, and today carrying his canteen inside his haversack instead hanging from his belt so that every time he stopped for a drink it took ten times longer than anyone else and they had to stop and wait. Faced with their irritation and getting tired, Mickey fell further back, and 'focussed his attention on the counsellors (usually myself)', Carper wrote.

They hiked across a meadow, the grass swishing and clouds of midges rising at their approach, with Kelman and John far ahead and the boys strung out in clusters of twos and threes, Mickey and Carper at the tail end. When they stopped beneath a massive spruce tree to rest in the shade, Kelman and Carper quickly conferred. To keep the boys together in a single group, Kelman gave John, who kept racing ahead, a large can of beans to carry to slow him down, and Carper shouldered Mickey's backpack so he could more easily keep pace. There were certainly scientific reasons why they wouldn't want to lose a subject from the experiment, but I wondered if I read sympathy in Carper's notes, as if he could identify with a boy who felt like an outsider.

The day hike was the first of a whole range of group activities Sherif had scheduled for the next five days and the second stage of the experiment. He predicted that at the end of that time, the boys would identify closely with their new group, and would experience a 'feeling of belongingness'. They would have a clear leader and express their collective identity in shared catchwords, group slogans, and ways of policing rivalry or friction between group members, all visible signs of what he called group norms. It was as though the boys would shed one identity and a new one would take its place. Sherif wrote that for each boy, 'His sense of personal identity does not and cannot exist independent of the group setting. In short, the individual cannot be considered apart from or in contrast to the groups of which he is a member … the tired but still popular question concerning the individual versus the group … simply evaporates into thin air.'

By the time they got back to camp later that day, sweaty and tired, the men leading both groups had been fielding questions from the boys about their friends in the other group: what they were doing and when they were going to see them. The written instructions to staff were silent on how to handle the boys' curiosity. It does not seem to have factored into their preparations. Sherif had Sussman instruct the men to tell them the 'camp was organized to try out different camp activities … they were engaged in different activities and were not to be interrupted'. But this didn't satisfy the boys and, back a camp, a now-glum Laurence and the other boys in Carper's group hung around the kitchen and pestered the cook for an explanation. But he said he knew as little as they did. Sherif had Sussman revise the daily schedules and arrange for meals to be served at different times so the groups didn't cross paths and there would be no contact between them, so it was likely the cook knew a little more than he was letting on.

At different times that afternoon, the adults organised a treasure hunt for each group, to build morale and to get them working as teams. I imagined the boys absorbed in the task, hurrying from clue to clue, in and out of the shade of the woods, down by the stream, to the mess hall, their tent, and back again, as the men silently shadowed them, noting any boys who seemed to take the lead, and which ones seemed happy to follow. Both boys won a $10 prize, as Sherif and Sussman had pre-arranged and noted in the day's notes.

The second test of how the group was developing was

how they made decisions about spending the prize money. Before supper, Jim Carper crept up to his group's tent and listened outside as the boys discussed what they should buy with their money. The light was fading; one of them jiggled a torch, and shadows jumped and fell on the wall of the tent. 'Hey,' someone said, 'don't waste the battery.' I imagine it was John, the practical one, who suggested they use the money to buy a Coleman lamp so they could play games after dark. Maybe it was Walt, one of a family of six boys, who objected that it would be impossible to divide up a lamp once the camp was over, that they should simply divide the money between them.

'Rubbers, we should buy rubbers!' one boy said. This sounded like Mickey to me, eager for attention.

'You've got a head full of rubbers,' another one said dismissively.

'We could get a girl for the counsellors, with a conveyor belt so they don't have to walk out of the tent,' the boy I guessed was Mickey said loudly. 'A conveyor belt with a whore at the end of it!'

There was a moment of shocked silence and then a jumble of annoyed voices drowning Mickey out. I imagined Carper straining to identify who was speaking in the excited muddle of voices that followed, suggesting that they use the money to rent a canoe, buy a horse, even buy a hot rod. I imagined his mounting impatience as he listened to this scene and the boys' inability to settle and make a serious decision, offering instead a raft of increasingly 'unrealistic suggestions'. How was he supposed to observe who exactly was making suggestions and how they were being received when he was out here in the dark, the mosquitoes whining, tired from the day's exertion and the prospect of another long night ahead typing up notes before any possibility of going to sleep? And the boys clamouring, calling over one another, offering increasingly half-baked suggestions, no particular boy demonstrating that he had any more influence than the rest. According to Sherif, by now one boy should be taking charge and directing decisions.

But $10 was a huge amount to the boys, worth the equivalent of almost $100 today. In 1953, $10 would buy one hundred sodas, seventy hotdogs, or a week of canoe hire. The amount of money had unsettled them; it seemed excessive, and some were feeling uneasy about the way they'd 'won' it. Perhaps the prize money had been a test. They were also still bothered by the mystery of Harry Ness's about-face and worried that he'd separated them from their friends as punishment. Maybe they could use the money to buy their way back into favour.

'We should give some money to the counsellors.'

'— five dollars to Mr Ness.'

'Did you see after he gave it to us he walked away real quick?' I imagine this might have been Harold, who paid such close attention to the men.

'He didn't say grace today. He always says grace.'

'He's mad at us?'

'He split us up, didn't he?' a boy said impatiently.

'But what did we do?'

There was a short silence.

'We did something,' another boy said glumly.

Their conversation was cut short by the clanging of the dinner bell.

While Carper's group had come to no decision, Jack White's group had decided after some discussion to buy something that could stay on at the camp after they'd gone. They opted for two flags — an American flag and a camp flag with an emblem on it. But what emblem? An eagle? A wolf? Under the sway of the ghost stories they'd been telling the night before, one boy suggested a picture of a wolf eating a gallbladder, but the others dismissed the idea. Perhaps under the influence of their jungle comic books, they finally decided that the second flag should decorated with the image of a black panther.

Once the table had been cleared, Carper announced that the caretaker was compiling a list of supplies for his morning trip into town: what did they want him to buy with their prize money? The boys exchanged glances and then John piped up that they'd decided on a Coleman lamp. Were they sure? Carper asked, glancing from boy to boy around the table, remembering the arguing that had gone on just an hour before. Was it a unanimous decision, Carper wanted to know. John turned to the rest of the group and asked for a show of hands. Sherif, hiding in the kitchen, would likely have watched impatiently, irritated by Carper's questions. It was enough that the boys had fallen in with John's suggestion. It showed a hierarchy had developed, with John as a leader whose plans the others followed. Still, Carper looked doubtfully at the group of boys, worrying that the quieter ones had been railroaded. Sherif hurried into the mess hall, in my imagination pulling out a small notebook and the stub of a pencil from his pocket, like an actor hurrying onto a stage. He licked the stub and flipped the notebook open, and in his heavy accent announced theatrically, 'One Coleman lamp, coming up.'

Sunday morning was overcast, the air thick and muggy. The boys in Jack White's group had rushed to the mess hall at breakfast time, expecting to see the others. But when they got there, the cook was wiping the second table with a wet rag and it was clear the other group had been and gone. The room would have felt empty and quiet with just the twelve of them there. It was beginning to dawn on these boys, like Doug and Peter and Irving, that they were not just in separate tents but were being kept apart. Peter, who had relished the archery games with his friend Laurence, was feeling resentful, and he called out sarcastically to the cook in the kitchen, 'Hey Sandy, why don't we eat together? Not allowed to see our friends, eh?' But Sandy just shook his head at the boys, saying nothing.

After breakfast, while the others returned to their tent, Peter, Irving, and a boy called Nathan ran quickly in the other direction, towards their friends' tent, hiding behind bushes and trees, keeping out of sight of the camp counsellors. But Jack White spotted them as they burst from some bushes and ran low across the stream, and he ran after them. In the last 100 yards, they called to their friends, 'Laurence! Eric! Hey!' The tent flap burst open and a group of boys spilled out, whooping hellos, but Jack White had caught up to the three runaways, blocking their path and saying that Mr Ness had insisted they return to the mess hall. He herded the reluctant boys away. For Irving, it had been a daring prank, and he was excited and cheered by the adventure. But Peter walked sullenly ahead of Jack White, saying nothing, while Nathan kicked at the dirt, raising dust.

Sunday church services, along with early morning reveille, were part of summer-camp tradition. But a visiting minister might ask questions or cause problems, so Sussman volunteered to fulfil these duties, eager to take a more active role in the study rather than being caught up in the daily chores of arranging food, supplies, and maintenance. He'd volunteered his services for this experiment because he felt under-appreciated at Union College, where at his first faculty meeting after his appointment as assistant professor in 1951, he was puzzled when the president introduced him not by his new job title but as coming from a long line of watchmakers with skills in restoration and repair. To his dismay, he discovered that five of the faculty — including the dean — were avid collectors of antique clocks. He spent most nights working late at his workbench, making repairs and trying to ignore his growing humiliation and the conviction that it was these vocational skills rather than his academic qualifications that had gotten him the job. Now here he was in a similar situation. Instead of being treated as an equal in the research project, he was stuck doing the menial work.

That afternoon, Sunday 26 July, each group cleared a separate space in the woods and fashioned an altar. Sussman took charge of the proceedings, and held two different services, complete with hymns and a homily on the importance of 'clean minds'. The choice of topic was no coincidence. Both White and Carper had reported uncharacteristic swearing and dirty talk in both their groups. If Sussman preached against cursing and the boys continued to do it, it would demonstrate the strength and power of their 'emerging group norms'.

Sussman's sermon clearly touched a nerve because after the service someone from White's group, Brian Kendall, a tall boy whose smile revealed a large gap between his two front teeth, came to see Jack White, looking troubled. Biting his lip, he confided to Jack that the night before, some of the boys had formed a swearing club, with an initiation ceremony. He told Jack that he didn't believe in swearing and he didn't like it, clearly expecting Jack White to intervene and put a stop to it.

'That would have been me all right,' Brian laughed on the phone. 'I remember that in our group there were a few boys who just kept pushing the boundaries. Kids will do that,' he said.

Brian's conversation was peppered with these easy, authoritative statements about children. He had spent his working life as a junior-high teacher. When I spoke to him, he was retired. At the time of his conversation with Jack White, he didn't yet know the camp was an experiment. But White's response should have tipped him off. 'He didn't react the way you'd expect,' Brian said. 'You've got to remember even "damn" and "shit" were shocking words back then, and we never, ever used the f-word. This kind of language really was taboo. It wasn't just vulgar — we were taught it was sinful.'

Brian's memories of the camp were sketchy. He remembered little about the other boys, but he did remember feeling guilty about telling Jack White about the swearing club. But if he was worried about getting others into trouble, he needn't have been. White had no intention of putting a stop to it. He was more interested in finding out from Brian exactly what was involved in the initiation ceremony — 'saying a sentence of 6 words, of which 5 were profane' — so he could describe this exciting development, proof that the group was developing its own rules, to Sherif.

But given how taboo Brian said swearing was, I was curious

about just how it had started. How did it happen that boys who had refused to sing the word 'hell' on the first night of the camp because it was curse word had ended up just a few days later forming a swearing club? Perhaps Brian was right, that some boys were seeing how far the men would allow them to go. And they read the lack of censure as a form of approval.

Yet there's a clue in the notes that the some of the men were actively encouraging it. OJ Harvey seemed to have a fluid role in the camp — disguised as assistant camp manager, he had free rein to drop in on either group on any number of pretexts, from asking what items boys wanted in stock in the camp store to collecting their mail. Harvey had been 'initiated' into the swearing club, earning membership by demonstrating his skill with foul language. No wonder Brian was bothered and approached Jack White after the Sunday service. But White's response didn't make him feel any better. 'I was left with a kind of dirty feeling,' Brian said on the phone. 'Looking back, I guess I thought I was doing the right thing, but it didn't feel like it. I probably sensed then that we couldn't rely on those guys.'

In the background, classical musical swelled from a radio. There was a long pause.

Brian cleared his throat. 'You know, I'm a volunteer at a community garden. A lot of city people, you know, don't know much about gardening, and that's part of what I do, I show them how. When I was growing up, my mother raised chickens and grew vegetables, but she could never get me interested, you know. I always had my head in a book. But I remember when I got home from that camp, the first thing I wanted to do was go out with my mother and feed the chickens and look for eggs.'

Brian laughed suddenly, like a trumpet blast. 'I wonder what my mother made of that!'

At the time I was thrown by this seemingly abrupt change of subject. But later it occurred to me that Brian was connecting his interest in self-sufficiency to something that happened on the camp. Perhaps White's failure to intervene and put a stop to the swearing had taught Brian a powerful lesson: don't turn to others to look after you; you have to learn to do it yourself.

By Monday morning, the effort of keeping the two groups apart was proving too difficult to sustain. Sherif decided that getting more physical distance between the groups was a priority. He scheduled a three-night camping trip for each group, starting on different days and travelling in opposite directions.

The first group to leave was Carper's, on Tuesday morning, and from his description of the boys it's clear they were relieved to be getting away. They hadn't been able to shake the feeling that Harry Ness was punishing them by keeping them from their friends. Ness was staying behind at the camp, so this trip meant they could forget about him for a while. They packed their new Coleman lamp carefully, then hoisted rucksacks and tents into the back of the truck and pulled over the tarpaulin, chattering excitedly as the truck bumped away from camp in the direction of Lake George. By the time they arrived, their worries about Ness and sense of guilt and unease from the night before seemed to have dissipated. They tumbled out of the truck and walked the last two miles to a campsite on a long finger of land that jutted out into the lake. In contrast to the deep green Camp Talualac, enclosed by woods, here their campsite overlooked the vista of the blue lake dotted with islands. The breeze off the lake was warm, but still a change from the muggy closeness of the camp in the woods.

After lunch, the boys took the small track down to the end of the point and round to a sandy bay, where they played in the water and John showed the others how to build a dam to trap fish, then supervised its construction. Mickey threw himself into the task, and even Harold, who had a habit of wandering off on his own, joined in, collecting rocks and building an elaborate wall to stop the rushing water and make a pond. Laurence seemed to have cheered up and taught the others to make farting noises with their wet armpits. I pictured Carper sitting on the beach, leaning casually back on his elbows, chewing a blade of grass, as if enjoying the view of the thickly wooded mountains rising on the other side of the lake. Through half-closed eyes, he watched the activities of different boys, and memorised snippets of conversation, as well as committing to memory who was giving instructions and who was carrying them out. Carper's notes described not a single group of children but pairs, trios, and loners. Even so, they seemed to be content. I suspect that Jim Carper was feeling more relaxed too, free of the scrutiny of the intense and uptight duo of Sussman and Sherif. Reading Carper's notes from this trip, it was the first time I began to wonder how the mood of the men may have affected the children and whether the push–pull of the men's emotions could be mirrored in the boys' interactions with one another.

While Carper watched the boys, I imagine he had time to think about the distance between the rules they'd agreed to and the reality. Sherif was eager for him to speed things up a bit, feeling that unanimous group votes on activities was slowing things down. But Carper's notes hint he was uncomfortable with the idea. It was true that the research predicted a hierarchy would develop, but the idea that one or two boys be allowed to hold all the power and wield it over the rest didn't sit right with Carper, and he stuck with his practice of encouraging group decisions. That night in his notes, he demonstrated his resolve to stick as scrupulously as he could to Kelman's instructions. He described how neither he nor Rupe, the junior counsellor, took a leadership role or made an attempt to influence the boys that day. They didn't help the boys in preparing and cooking their food or in pitching their tent, even though the boys were tired from the afternoon of swimming and struggled to put up the large canvas tent on their own. Harold yelled at Carper, 'We need somebody's advice.' Carper wrote, 'This was directed to the counsellors, who said not a word.' But the impracticality of non-intervention was obvious to him too. He confessed in his notes that he had to 'goad' the boys into getting supper ready and set rules about when they went swimming.

While Carper was trying to maintain an arm's length, back at base camp it seemed Sherif had abandoned this idea altogether. Straying from the research team's agreement to shadow the boys in secret, Sherif had begun working enthusiastically alongside them on a series of chores around the campsite. Instead of standing back to watch how the boys made decisions and letting the group form naturally, Sherif seemed intent on moving things along. On Monday, they had to build a latrine, erect a hut, and dig out and remove a large boulder from the hut's floor — all of which required the combined physical strength of the whole group and sometimes the assistance of Sherif-as-caretaker and

Jack White as well. But the men only pretended to help, secretly pulling instead of pushing to make the boys work together all the more, and forcing one of them to take charge and give directions. I thought of Doug, a self-confessed puny kid, lifting and hauling rocks, and Brian the homebody, who loved reading, and how this labour sounded more like something to be expected of men in military training than eleven-year-old boys.

But as far as Sherif was concerned, whether they were a collection of children or an assemblage of men didn't matter: groups as large as 'ethnic groups or nations' or as small as 'a fraternity or sorority or a well-knit club' shared the same properties. What Sherif apparently couldn't see, or wouldn't acknowledge, was the particularities of this group of boys that undermined this larger narrative and limited any generalisations he might like to make. They were children in an alien environment, surrounded by adults whose behaviour puzzled and sometimes troubled them, and they were far from home.

On Tuesday morning, just before they left for their camping trip, Sherif-as-caretaker came back from town with the flags the boys had ordered. They gathered around excitedly to unpack them. First they unveiled the American flag and were overawed at its large size. Before they could undo the other parcels, Jack White told them the story that Sherif had concocted that morning. Unfortunately when Sherif got to the store, he discovered the order for a Panther flag would not be ready before the end of camp, so instead Sherif had bought a plain flag that Jack could help the boys decorate. The good news was that the store had offered them a 'special purchase' — for fifty cents, each boy could have a t-shirt and a matching cap. But it was an 'all or nothing' deal — every boy had to buy them or the deal was off. I was surprised when I read this. Presumably Sherif had dreamt up this excuse so that he could demonstrate that the boys had created their own 'group products', but it was direct manipulation of the situation to a level I had not encountered in the notes before this point. It made me wonder if spending the treasure-hunt money on the flags had been entirely the boys' idea after all, or if Sherif or Sussman had played a role in planting that notion as well.

A few boys made a start on decorating the plain flag, with camp manager Mr Sussman helpfully providing a picture of a panther's head to use. But drawing on the fabric was difficult, so Jack White did most of the work. While the boys played a game of dodgeball, White stencilled the panther head onto the flag. Was it a spontaneous decision by the boys to have the word 'Panthers' stencilled on their t-shirts too? White's notes certainly make it appear that way. Once one boy had it done, the others clamoured for the same design.

By mid-morning, when Jack White left with the boys for their camping trip to a spot near Sacandaga Reservoir, all of them, including Jack White and Ken Pirro, the junior counsellor, were wearing their uniform of Panther t-shirts and caps. They carried with them their new Panther flag. But was it deliberate or a mistake that White had left a critical detail the boys had requested — olive branches, symbolising peace — off the design? The flag flapped in the hot breeze as they hiked; the panther's mouth snarled, its sharp black teeth silhouetted against the white background.

They might have had a group name and a flag, but at times there was little sense of camaraderie. White's Panthers were a divided bunch. On one side were boys like Peter and Brian, and

I'm guessing Doug too, who had protested about the swearing club. On the other side were boys like Nathan and Joe, who instituted 'depantsing', where they pulled the trousers off boys and threw them up into a tree as punishment for slowing them down or failing to do their share of chores. Jack White excitedly described this 'corrective' as a sign that the group was policing itself. Correctives were 'examples of censure or punishment of boys either by other individual boys or by a whole group. Examples included being ignored, ridiculed, chided mildly, berated, physically punished.' White, like Sherif, read the instigation of punishment as a healthy sign of the development of the group's shared rules, ignoring the fact that some of the boys objected to it, and missing an alternative explanation — it was the behaviour of unhappy children.

After the long hike, White's Panthers were relieved when they reached their campsite and excited at the sight of water glinting through the trees. They dumped their belongings and raced straight to the inlet. It was a shaded swimming spot where a forest stream widened and flowed over a jumble of flat rocks before emptying itself into the lake.

After the swim, the boys were hungry, but the truck with their food supplies hadn't arrived, so they set up their tents and collected rocks to make a fire ring. With still no food in sight, White wrote in his notes that two boys suggested that to pass the time they should play strip poker. In brackets, White added for Sherif's benefit,

> (All of the boys are very careful not to disrobe in the presence of others. Most of them put their pajamas on under their

> blankets after getting in bed. Playing strip poker is considered quite daring, and one of the penalties instituted by the boys for losing is that the loser must do a hula dance in the nude when he is completely stripped. XX [name removed] informed the P.O. [participant observer] that he was forced to do the hula when he lost recently).

I did a double take when I read this. White's casual tone implied that this was not the first time the boys had played this game. The contrast between their modesty, getting changed under the cover of blankets, and the image of a naked boy dancing in front of the others disturbed me. Who had instigated the game, and how had the boys' seemingly powerful taboo against nakedness been reversed, and in such a short time? For all Sherif's neutral language about group norms 'arising spontaneously' in his proposals about this research, this shift in the boys' inhibitions seemed out of character. The dirty talk, the initiations, the humiliation games sounded the kind of antics and hazing that happened in frat houses or army barracks, rather than among ten- and eleven-year-old boys. Or was I being naive? I called Brian, I emailed Doug.

Brian didn't remember much, but guessed it could have been just talk, more evidence of the boys' attempts to get the adults to take charge. 'We kept waiting for the counsellors to react, to do something. To a kid like me, who liked rules and who was always out to please, it was a pretty uncomfortable situation.'

Doug didn't recall it either. 'That almost sounds too bizarre even for those clowns — in 1953 we wouldn't have even known swear words or understood strip poker. *If* that had happened, it would have been awful for boys of that era.'

Was Jack White a reliable narrator, I wondered. Perhaps as a graduate student whose dissertation relied on this experiment, White was consciously or unconsciously shaping his notes to please Sherif rather than to reflect reality. I compared Herb Kelman's observations of the group on that same afternoon. There is no mention of strip poker. Kelman described the boys' pride in their Panther flag, and how he watched them playing by the stream, where Irving took off his shirt, 'revealing his Panther t-shirt underneath. He said: "This is a job for Super Panther!"' Here Kelman added an explanatory note in brackets, thinking that Sherif might not have been familiar with superheroes: '(imitating Superman, who always takes off his regular clothes and reveals his Superman outfit before getting into action).' A little later, Kelman wrote, the boy put his shirt on again and said: 'Now I'll get back into my disguise', to the cheers of the other boys. Could both accounts be true — the mini-men playing strip poker and the innocent boys playing at superheroes?

I could guess which account Sherif favoured. It was day five of this second stage, and White's group had a name, a flag, a uniform. Kelman's observation pointed to a benign shared group identity, but White's took it even further. Strip poker was proof that the norms of the group were so powerful they overrode individual conscience.

At the lake, the truck finally arrived and the Panther group helped unload it and began cooking lunch. The air filled with the smell of woodsmoke and hamburgers. After lunch, the sky clouded over and the lake was grey and choppy. They spent the rest of the afternoon at a nearby airfield, where the boys played on an abandoned B-24 bomber, pretending to fly over enemy territory. Irving claimed the captain's spot in the cockpit and refused to move, arguing that his uncle was a decorated fighter pilot so he was best qualified to be captain. In contrast, Doug took turns with the other boys in the co-pilot seat and tried to persuade Irving to be fair. Peter stood below the cockpit with his hands on his hips and yelled up at Irving, who ignored him. Peter complained to the two counsellors that Irving wouldn't give anyone else a turn, but Jack White waved him away, telling him it would be fine. By the time they rounded the boys up for the trip back to camp, no one waited for Irving, who was the last to scramble down from the plane and he had to run to catch up with the others.

Back at camp, Irving flopped down by the campfire, which Peter was fanning as Brian lay twigs on the embers to try to get it going. The wood was damp and smoke poured from the fire. Irving turned his head away, squeezing his eyes against the smoke and fanning at his face. 'Hey!' he complained.

'You could get up and help,' Peter said. 'You've sat down all afternoon!'

'I want to do the cooking.' Irving coughed and wiped at his eyes. 'You guys never let me.'

'That's because you burn everything!' Peter said. 'You say you'll do something and then you walk away and leave it to someone else.'

'I want to cook,' Irving said. 'You can do the dishes for a change.'

'I'm cooking!' Nathan said angrily to anyone who'd listen. He pointed at Irving. 'You look after the fire.'

'We'll make a roster — we'll list all the jobs and make sure everyone does their share,' Peter said. 'Go and get a pen and paper,' he told Nathan.

'You're not the boss!' Nathan took a step towards Peter.

'Yeah,' Irving muttered.

Peter shrugged, threw the large leaf that he had been waving at the embers into the fire, then sat down and crossed his arms.

'You!' Nathan turned on Irving, who stood up quickly and began to back away. Nathan lunged at him and threw him to the ground. 'This is for bragging all afternoon about your "war hero" uncle!' Nathan yelled, tugging at the zip on Irving's pants. Irving struggled and yelled. Boys came running from the woods. Some ran and piled on top of Irving. Others, like Brian, looked to Peter, waiting for him to intervene, which he usually did. But Peter frowned into the fire as if he couldn't hear a thing. Irving was gasping for air and sobbing so loudly that Jack White came out from his hiding place. 'I had to call the boys off,' he wrote to Sherif.

Once freed, Irving, whose Panther t-shirt was covered in dirt, ran away to the bushes and cried. White sat on a log at the campfire and tried to draw the boys into a discussion about preparing supper, but Peter sulked and refused to be drawn and the others were subdued. Until now, Peter had been White's pick as the leader who would take control and set boundaries for group behaviour, but his domineering manner irritated some of the boys, and he brooded when things didn't go his way. White looked around impatiently, mentally crossing them off — Nathan, who was hacking away at a log with a hatchet, was too aggressive; Brian, who glanced worriedly at the silent Peter, was too weak. Then there was Doug, who moved around the edge of the clearing, collecting firewood. White had noticed him that afternoon too, and his knack for getting the others to take turns playing on the plane. As White watched, Doug turned and called, 'Who's gonna help me with this wood?' balancing a pile in his arms, and three boys, including Nathan and Brian, darted forward to help him. Yes, White thought, even though he was one of the smallest boys, Doug Griset was a clear contender as group leader.

The trajectory of the Panthers was beginning to mirror that in Golding's novel, where, after the initial chaos and confusion, the boys turned to a leader who was constructive and fair and who seemed to provide a sense of security in a shifting and uncertain situation.

Meanwhile, 30 miles to the north, two nights into their camping trip at Lake George, Jim Carper's boys woke up to the honking of ducks on the lake as the sun came up on Thursday 30 July.

Compared to the Panthers, Carper's group seemed a relatively happy bunch, but there was still no clear leader taking charge. Most of the boys wanted to cook, so mealtimes were chaotic, with two fires going and boys jostling and no one coordinating. Hamburgers were fried and ready to eat before potatoes were even peeled, some boys were lining up for third helpings before others had had their first. John, who had plenty of camping experience, tried to give advice, but he wasn't assertive enough and the others wouldn't listen. Even Laurence, who was good at coaxing the others to do things with a laugh and a joke, gave up trying to get the boys to line up and take turns when it came to cooking and serving.

That morning, Carper decided he had to take action. None of the boys had washed the supper dishes from the night before, and there were plenty of animals in the Adirondacks, including bears and coyotes, who could be drawn to food scraps. Carper used passive voice in his observation notes to distance himself from directing the boys: they were, he wrote, 'presented with an agenda', which was '[s]tart fire, wash dishes, make breakfast, decide when they want to go home'. In response, the boys organised themselves efficiently, and after breakfast of hot chocolate and bacon and a thorough camp clean-up, the boys had a group meeting to decide the day's activities and menu. Carper watched closely, waiting for a leader to emerge, but none did, and the chaotic-sounding meeting that he described in his notes took over forty-five minutes, with boys wandering in and out and shouting non sequiturs as the men looked on. If it was frustrating for the men, it was strange to the boys, who expected the adults to take the lead. Walt remembers this as one source of his uneasiness about the camp. 'There was no one in charge saying, "We're going swimming; we're going boating; we're going fishing." It was very unstructured. We had a lot of time to do things on our own, we could do whatever we wanted.'

What's striking about Carper's notes is how he followed up with boys who weren't participating, ones who he described as 'on the periphery'. It wasn't part of his scientific brief but a mark of his concern that each boy felt included. During the morning's protracted and often rowdy meeting, he had sat with Mickey on the edge of the group and encouraged him to make suggestions. Later that afternoon, when the boys were swimming, Carper noticed that Laurence, who was usually in the thick of group activities, stayed back at camp and sat by himself at the campfire. When Carper asked him what he was doing, Laurence showed him how he was whittling a piece of wood that would be big enough to fit the names of all twenty-four boys on the camp. Carper paid attention to individual boys and described in his notes some who spent time on their own or thinking about their friends back at camp. Probably because he had little contact with the research team, Carper had little idea of the impact these kinds of details could have for Sherif and Sussman; he was simply reporting what he saw. In contrast, it seemed that White intuitively understood that his observation notes were meant to reassure Sherif that the experiment was going to plan.

For, back at the main campsite, Sherif was likely on tenterhooks. With both groups gone, he could not observe them for himself, and he waited impatiently for the sound of Harvey returning in the truck in the evenings with the men's daily reports. He and Sussman had plenty to do in the boys' absence — there were the preparations for the competitions to finalise, the planning of scenarios and the writing of scripts to make the tournament announcement seem natural. But I imagine he would have found it hard to settle.

On their final night at Sacandaga, the Panthers decided they wanted hotdogs for dinner because they were quick and easy to cook. But knowing how much his professor liked to double-check every small detail, White called Sherif from the store to get his approval for the evening's menu. The boys were enthusiastic about cooking hotdogs on sticks over the fire, but White worried that Sherif would object because this wouldn't involve any group work and he would have to come up with an excuse and an alternative for the disappointed campers. To his relief, Sherif agreed to the purchase, and White stocked up and headed back to Sacandaga. When he arrived back, the boys had built a fire with pine needles and logs, so they were soon cooking hotdogs over the fire, the fat spitting in the flames.

After dinner, talk around the campfire turned to extrasensory perception. The boys took turns to see if they could read one another's minds. One boy put his hands over his eyes while the other stood around 20 feet away, looked up at the sky, thinking of a number. Brian told the group how he and his mother could often read each other's thoughts, and White asked him what his mother was thinking now. Brian said it was too far, thoughts couldn't travel that distance. Whether it was Brian's talk of his closeness to his mother that conjured the image for each boy of his own mother, or the lights of High Rock Lodge that twinkled across the lake that reminded the boys of home, I don't know. But later, in the tent, an argument broke out about the swearing club, and some boys threatened to go home unless it stopped: White wrote in his notes that Brian called out to him and asked him what time the next train left for Schenectady, and Peter, too, was loud in his protests. The arguing and the thoughts of home had stirred some of the boys, and sleep did not come easily.

Those who were still awake an hour later would have heard the truck rumble into camp. White's phone call that evening had agitated Sherif. After he hung up, Sherif began worrying that allowing the boys to cook solo meals had been a mistake. With no means of contacting White, Sherif insisted that Sussman drive to the campsite, using the excuse that he had forgotten to supply groundsheets for the boys' tent. But by the time they got there at

8.30 pm, the boys had already turned in.

Soon after, Doug began complaining of a stomach-ache. During the night he got up three times and woke White, saying that he felt sick. But with nothing in the first-aid kit to help him, White told Doug each time to go back to bed and wait till morning. The fourth time that Doug woke him, White told the junior counsellor to take Doug back to the dispensary at the camp. At 4.30 am, Doug was 'marched through the woods' for an hour, arriving back at camp at 5.30. White's use of the word 'marched' sounds like a rebuke, as if the boy was being weak, suggesting White had little sympathy. Or perhaps it was that White anticipated that Sherif would be angry and blame him for allowing Doug to be returned to camp, so he made it clear in his notes that he wasn't being soft on the boy.

Doug doesn't recall the walk, but we both marvel at the image of his ten-year-old self hiking through the woods in the dark, the faint glimmer of the junior counsellor's torch dancing ahead of him, his heart thumping, perhaps, at the scream of an owl, at the crackle of twigs and leaves that could signal an unseen animal in the undergrowth.

On Friday, their final morning at Lake George, Jim Carper and his junior counsellor took charge, cooking breakfast and delegating jobs so the boys would be packed up and ready in time for the truck's arrival. On the way back to base camp, Carper wrote that all the boys were engaged in 'individual activity' — chewing candy, humming, and gazing out the back of the truck as it bumped along the dirt roads. Back at camp, Harold went to the infirmary with a 'fever', joining Doug from the other group, whom the nurse had already put to bed.

Sherif might have chosen Mrs Terani, the nurse, because he thought she would be no distraction to the men, but the infirmary seemed to be a magnet for some of the boys. The men's notes from this time onwards are dotted with references to boys missing from mealtimes or games because they had gone to see the nurse about 'medical issues'. I couldn't work out if it was because the boys were genuinely ill, or if they went there for sympathy and comfort. For some boys, perhaps it was easier to say they were ill than to admit that they were longing for home.

After they'd unpacked their gear, Carper called in to the infirmary to see how Harold was feeling. The boy was sitting up in bed when Carper arrived. Harold was the kind of boy used to working things out, who looked closely at things to figure out the logic behind them. But every time he asked the adults questions, he could never get a straight answer. From that first day, when he'd asked Carper about the microphones in the mess hall, he had the sense they were keeping something back. When Carper sat down on the edge of the bed, Harold told him he'd worked out it was important to the men that the boys make friends and get along in their new groups. He didn't bother lying: he told the counsellor that nothing was really wrong with him, but he wanted to go home. Carper wrote, 'He vowed I would not change his mind and that if we did not let him go home he would wreck the group. He prides himself on knowing everything that goes on around the camp.'

When Carper saw how determined the boy was, he gave up trying to persuade him. 'I told him I would discuss it with Sussman. He said that Sussman would not let him go home and

I should pass his threat along to Sussman. That evening we called his parents and although they were not eager for him to go home plans were made for him to leave the following day.'

Doug, who was in the room next door to Harold, was homesick too. While Doug doesn't remember many details of the camp, he's certain of this one thing. 'Oh, I would have been homesick all right. I would have been homesick to beat the band,' he told me. 'You know, I'm speculating here, but you can see why they might not have wanted the parents to visit. All the kids would have been blubbering and saying, "I want to go home."'

Doug's homesickness triggered anxiety among the researchers. White and Sussman worried that because Doug was a popular boy and a potential leader, others might follow his lead. Already two Panthers, Peter and Brian, had threatened to go home. Sherif wrote in his notebook that White and Sussman were 'gloomy' about Doug's 'ills' and 'feared contagion'. They were seven days into the second stage, and the hypothesis that each group would share a sense of camaraderie and feel protective of its members, would have a clear leader and shared ways of doing things, seemed further away than ever. With the two groups failing to bond, and the risk of homesickness ruining their plans, Sussman and White argued for starting the next stage of competition between the groups. But Sherif agonised over whether to give the groups more time to bond. He worried aloud that the two groups were still too fragile; perhaps it was too soon. But the others argued that competitions and the prizes would ignite the boys' enthusiasm and make them work as team. Finally Sherif was persuaded. He scrawled in his notebook that White and Sussman 'urged me to start stage 3'.

OJ Harvey wrote the next day, 'Several Panthers want to go home so [we] brought stage 3 forward.' The next stage, the men hoped, would rally the boys into working teams. And at first it seemed to work.

# 6
# Showdown

Harry Ness tugged the pencil out from behind his ear as he waited for the Panthers to finish clearing away their breakfast dishes. The script had been typed the night before, but that morning Sherif had revised it again, so Ness had had no time to practise it.

In the kitchen, someone ran cold water into the frypan of bacon fat and hissing steam rose in a cloud. The boys jostled and jiggled at the table. Doug Griset had been coaxed out of the infirmary for breakfast, and the others were glad to see him.

Jack White called, 'Listen here, Mr Ness has something to say, y'all.'

Ness cleared his throat and read from the clipboard, announcing that the other group of boys had challenged them to a three-day competition. The Panthers exclaimed excitedly, and he paused and waited for the chatter to die down. 'This includes ballgames and feats of skill, such as tent pitching and tug of war.' He turned to the shelf behind him. 'The members of the team with the most points will win' — here he tugged at a cloth that was covering something standing on the mantelpiece and glanced again at the clipboard — 'these beautiful and expensive stainless-steel knives.' Twelve knives fanned out in a half-circle on a stiff cardboard display stand. Ness shifted the stand so the silver caught the light and twinkled.

It was as though the air had been sucked out of the room. There was complete silence until someone breathed 'Whoa!', and then everyone was talking.

Ness raised his voice above the noise. 'All the members of the group who win this tournament will enjoy the pleasure of owning such a fine camp knife. Remember, no one can win a knife by himself. So pull and work together. The only way you can win a knife is for your group to win — you have to get the most points by the time the tournament is over. So, go to it, fellows, work together as a team!'

Doug held his breath, gazing at the knives.

'Only the winning team would get a prize,' he told me. 'There was no consolation prize. They paraded those knives at mealtimes. And this was not some little thing, this was a bowie knife that they were giving ten, eleven-year-old kids. Probably illegal to carry. And I wanted that jackknife so badly — every kid wanted it. Boy, believe me, did we want that knife. They were real good at getting us to want that prize. I've never been someone who's had a use for knives. I'm not a hunter or into weapons. It surprises me how much I wanted that knife.'

Sherif had used expensive knives as a reward in his 1949 experiment and knew they were a good choice. He wrote that the boys in his first experiment 'prayed for' and 'dreamed' about those knives.

Doug remembers how often his team checked the barometer

propped on the mantelpiece that showed each group's scores. 'There were a great many competitions that they put us through — and I put it that way because it reminded me of my years in basic training: "putting us through" rather than enjoying. They were testing us all the time as to whether we could be better than the others.'

Ness made exactly the same announcement to Jim Carper's group, but they responded with mixed feelings. His inadvertently stilted delivery convinced the boys that he was still angry with them and didn't want them to win. And their winning was even less likely given that Harold had gone home and they were one team member down. But they were excited to be seeing their friends again, and hurried off to the first game. When they arrived at the baseball field, they were surprised to see the other group kitted out in matching t-shirts and caps, with their Panther flag hanging from the backstop. Laurence made a joke of it, pointing at White, who also wore a Panther t-shirt, and saying, 'You got a new recruit, fellas?' But some, like Mickey, hung back. He told the others later that the men had outfitted the other boys because they were their favourite team.

Marv Sussman, umpiring, was eager for the game to start and clapped his hands to get the attention of the new arrivals. Who was the captain, he wanted to know, and what was their team name? The boys glanced round at one another. I imagined Sherif on the sidelines, perhaps holding a rake as a prop, watching closely. After a moment, Laurence volunteered for captain and then looked round at the others and said 'Eagles?' tentatively. It was not a group decision, Carper wrote.

Meanwhile, White had named Doug as catcher for the

Panthers — a role best occupied by the team leader. Doug remembered being mystified by the staff's choice. 'I *never* was the catcher. I always batted first because I was small and fast and I could be the first person to get on base. But for some unimaginable reason they made me catcher. I had to be the smallest kid there. Catchers are big burly kids, like a block of granite, who are going to catch all the balls and block the plate.'

But Doug was a good choice in terms of team morale. Despite the Panthers' outward signs of solidarity, before the game started Nathan and Joe threatened to depants Irving if he struck out during the game. Irving pushed his glasses nervously up his nose. Doug went over to Irving. 'Don't pay any attention to them, they just have to have someone to take it out on. They all like you, just as I do,' White observed Doug saying in his notes. But Irving didn't look reassured, and kept glancing worriedly at Nathan. Doug guaranteed Irving that he had the 'best batting position on the team and they were counting on him'.

Soon the game was underway, and the sounds of the crack of the ball against the bat, the whistles and catcalls of the boys, and their ragged cheers drifted from the pitch towards the shadows of the trees, where Sherif stood well back, out of the full glare of the sun. The game went well for the Panthers, who got five runs in the first inning. When it was Irving's turn to bat, he 'hit a single and was cheered and embraced by his teammates', White wrote. But then things took a dramatic turn.

Doug remembered, 'They had me be catcher and I've got no protection except for a glove. I'm a little twink of a kid and I'm in the firing line for every kid coming round third base trying to score. So here I am, and I go over to block the plate when this kid came for third base, and he ran right through me.' *Whoomp*, and Doug was knocked out cold. The kid who knocked him down was a quiet boy called Tony Gianelli, who was described as a 'low status' boy, a term they used for shy or reserved individuals who ranked down the group's hierarchy in terms of leadership potential. But Laurence and John and the rest of Carper's group knew that Tony would never hurt someone deliberately. Okay, he didn't say much and he wasn't sporty, and sometimes he wet the bed, but there was no way he would have done something like that on purpose. Tony wasn't a physically adventurous kid. He'd only come on this camp because his older brother, his hero, had convinced him, telling him how much fun he had on scout camps and jamborees.

Doug's teammates, the Panthers, raced across the field and gathered in a circle around Tony, accusing him of doing it on purpose. Tony had a shiny black pudding-bowl haircut that ended just above his prominent ears. His hands were balled into fists and his ears had turned red. He had a slow fuse, but it was lit now. OJ Harvey and Herb Kelman hurried over and lifted Doug between them, carrying him off the field. I imagined Sherif smoking furiously in the shade of the trees, exhilarated at this turn of events, to see his plans coming together.

Sussman blew his whistle and called the teams back to the game. But the Panthers were outraged that Sussman didn't censure Tony for knocking Doug over. Nathan sneered at Sussman, 'What, are they paying you to root for them?' When play resumed, the Panthers vowed to 'win the game for Doug' and took up their positions again. I guessed that Sussman did this deliberately. What better way to have the Panthers unite as a single group and forget all ideas of going home? Harvey noted their 'sarcasm and aggressive behavior': they were 'hollering, spitting, calling names', yelling 'cream puff, rubber ass, sons of bitches'. At first Harvey thought it was directed to the other team, but he was shocked to realise most of the Panthers' hostility was directed at staff, who they felt had let Tony off the hook. Nathan called the other team's junior counsellor, Rupe, 'an overgrown turd bender', a term that Harvey — who had been initiated into the Panthers' swearing club and knew their rules — noted was the Panthers' 'most vulgar term'. What was the reluctant swearer Brian feeling now, I wondered.

As captain of the opposing team, Laurence called encouragement to the other boys, but his heart wasn't in it. They were still upset for Tony, who glowered at the Panthers, and they were worried about Doug, and intimidated by the aggression and cursing of the Panther team. They played desultorily in the grinding heat until Harry Ness closed the game down and declared the Panthers the winners.

On the morning of the second day of the tournament, Carper's boys lost both the tug of war and tent pitching, and on the way back to the mess hall for lunch, they concluded that the Panthers were a better team and deserved their win, the animosity of yesterday forgotten. But the last thing Sherif wanted was for Carper's group to concede the tournament to the Panthers. If they 'accept defeat', he wrote, there was no chance of conflict. To lift their spirits, he sent Sussman to town to buy the boys a set of t-shirts as a surprise, and watched in the mess hall after lunch when Carper made a show of the mystery parcel, encouraging the boys to guess what was inside. They crowded round while he untied the string and exclaimed excitedly as he held up a t-shirt. With Carper's encouragement, they chose a name for themselves. Some voted for Cobras, but finally they decided on the Pythons. Walt Burkhard didn't know what a python was, but once Laurence explained it was like a boa constrictor, he embraced the name excitedly along with the others.

Doug scoffed when I reminded him about the names for each team and wondered who had suggested them in the first place. 'Why would you call kids Panthers and Pythons, two killing animals? And then have them fighting over jackknifes, which you could use to kill somebody? Why didn't they call the teams Panda Bears and Dolphins or something? See, it's the premise — they weren't researching it, they were trying to prove they were right!' Doug was by then on his own investigative trail. He had contacted the archivist in charge of the Sherif papers to ask for material that described him at the time. Doug was interrogating the story of the experiment, sifting through his memories to confirm how difficult it would have been to persuade American boys like him to fight one another.

The Pythons seemed buoyed by the men's attention. Carper was taking the lead with them, and they responded eagerly. Later that day, he wrote in his notes that he was gone for half an hour, and when he returned, all the Pythons had climbed on the roof of his cabin:

> I was fooling around with them a bit when they all decided to take me on. There was about a 15 minute tussle with everyone rolling round in the dust. At the end I was

> completely fatigued … At one point they were all in the counselors cabin and when they were finally put out with alot [sic] of difficulty they still hung around …'

It was a striking passage — not just because of the affection and playfulness it revealed between Carper and his group but also because it was so different from the dry style that placed the men as passive observers of the action. Now, with Sherif's approval, the men had to show they were keeping the boys happy and preventing the homesickness that had threatened the experiment. But didn't Sherif realise that allowing the men to be cheerleaders and motivators of the boys prevented the development of group dynamics that he had intended?

At night, Sherif and Sussman pored over the observers' notes. The agonising wait for something concrete to happen in line with his predictions was taking a toll on Sherif. He rarely went to bed before 2.00 am, and he was so wound up that he had trouble sleeping. Flouting staff policy, he began to drink whisky after the boys were asleep. But that just made things worse. From their experience with Sherif at the University of Oklahoma, Harvey and White already knew that when he was drinking, unless Carolyn was around to keep it in check, he became paranoid and mistrustful and blamed those around him when things weren't going to plan.

Sussman did what he could to dispel Sherif's suspicions. Luckily, he had a thick skin. It had been developed during the war when, as a conscientious objector, he was assigned to work as a hospital orderly, and he had endured abuse from staff and patients at a time when many regarded people like him as traitors, spies, and worse. He tried to calm Sherif, taking dictation and endless notes on what to say the next day to keep staff in order.

At first, Sussman might have taken pleasure in this role: the satisfaction of feeling needed and important when he had begun to feel some days like a real-life administrator of the camp rather than the research associate that Sherif had promised he'd be. He stayed up with Sherif most nights, doing what he could to reassure Sherif that the next day would go without a hitch. But soon the lack of sleep began getting to him too.

Carolyn wrote doggedly every two or three days, telling Sherif news of herself and the children and summarising any mail that had arrived. On his birthday she sent him a new sweater as a gift. But he didn't answer her letters. Finally, she wrote in a burst of exasperation — and probably sensing something was wrong — 'For goodness sake, let us hear from you!' But the world outside the campsite had disappeared for Sherif. The isolated camp in the woods had become his 'whole universe'. I imagine it was in the middle of one of these sleepless nights, after a few glasses of whisky, that he finally replied to her letters. In an undated and rambling note, he wrote that life at the camp was 'hectic' and demanding. 'There is so much to do, so many significant and insignificant item to attend [sic]. I have to think and think hard to put all of them together — that means a constant state of alertness, tension and perspective. It is hard on everybody and especially on me.' Here he drew an arrow to the margin of the page and added: 'I suppose I should not have expected any lighter load than this. It is not fair to expect persons who are working so hard here to coordinate smoothly their pieces in this whale of a project to which I grew up during the last twenty years with so much pain and sweat.'

During the day he prowled the camp, making last-minute changes and increasingly inserting himself in group activities so he could observe the boys with his own eyes instead of relying on the observations of the men.

In this third stage of the experiment, Sherif was aiming for outright conflict. When the Panthers found a large fly in their tent, their counsellor suggested they name it after the other team and burn it. OJ Harvey 'took photos of the cremation', and, to make sure he got good-enough photos, White wrote, another fly was caught and 'the burning was repeated'. But over lunch the same day — the first time the boys had eaten together since their separation more than a week ago — the observers noted the boys stopped at one another's tables for 'friendly chat'.

And Carper's Pythons, instead of being fired up about their losses, were demotivated. Rather than the hoped-for talk of anger and revenge against their opposing team, the Pythons felt outclassed. Without enmity and competition, the experiment couldn't progress to the next stage. Sherif had used what he called 'frustration exercises' successfully in his earlier study: incidents where the men secretly vandalised the property of one group so they would blame their opponents and retaliate. It had been the flint that sparked physical fights between the groups in his earlier study, and he decided to use it again now.

In the mess hall, hearing the dejection of the Pythons, Sherif decided it was time to up the ante. He waited until Carper and his boys left, and White brought his victorious Panthers in for their dinner. Perhaps he instructed Sussman to do it, perhaps he did it himself, but while the boys were eating, someone took a knife to the Panthers' tent on the other side of the clearing and cut the rope of their Panthers flag, pulled it to the ground, and stamped on it with muddy boots.

The boys were dismayed when they got back to the tent and found their flag trampled. Peter saw Mr Musee and Harry Ness passing nearby and called out, asking if they'd seen what happened. The two men came over. Sherif answered that he hadn't seen anyone cut the rope, but that the other group had been in the vicinity earlier. The boys showed Harry Ness the vandalised flag. Then Sherif intervened and 'asked Mr Ness if it might be possible to have the two sides discuss their complaints together'. It's a sign of Sherif's impatience that he interjected. Did any of the boys wonder why the camp caretaker — whose job was to chop wood, run errands, and keep the grounds and buildings clean — weighed in like this?

Reading the men's observation notes against Sherif's notebook and the published version of the experiments, at times it was as if Sherif chose only what he wanted to see in the observation notes. He had read the men's descriptions of the ballgame, with the boys' spitting, name calling, and physical bullying, and interpreted it as the boys in one team discriminating against and belittling the other. In Sherif's eyes, this was evidence a major hypothesis had been proven: '[g]roup members will prefer friends from within their (new) group … Subjects whose initial personal preferences are in the other group … will develop negative attitudes verging on enmity towards the outgroup …' But he seemed to ignore the evidence that all the men watching had noted in one way or another — the anger and disappointment of both teams directed at the adults involved in the games. Harvey wrote how the Panthers accused him of 'interference' in the play, and argued with the umpire over his decision. When Sussman called a Panther runner safe when he clearly wasn't, the Pythons yelled, 'Kill the umpire!' Carper's Pythons discussed with him at dinner that they felt the Panthers' counsellors were coaching them to victory. The adults' attempts to fan enmity by skewing the scores, first for one group, then for the other, backfired. Despite the division of the two groups and the competition, the boys shared a common view that the adults were playing favourites.

This was not the way the boys expected the men to behave — particularly not when it came to adjudicating at a baseball game. Baseball was a hugely popular sport and many of the boys would have been members of the Little League, which, in the wake of World War II, was being used as a vehicle for promoting the values of American democracy. Boys were taught that in Little League, rules were applied in a spirit of fairness, and if there was a dispute, adult umpires made even-handed and just judgements based on what they observed. 'This type of loyalty is the same thing we call good citizenship as applied to the city, that we call patriotism as applied to the country,' noted William J. Baker in *Playing with God: religion and modern sport*. Above all, boys were expected to learn how to win and lose graciously. In Little League, boys learned the value of 'good-natured' competition, where there was no lashing out, and no one left with hard feelings.

Sherif, it seemed, had made a major tactical error in conceiving of sport, and baseball in particular, as a metaphor for war, and that in losing a baseball game one team would turn on the other for revenge. Herb Kelman and Jim Carper wrote later that this was a major problem with the experiment:

'Here a cultural pattern of sportsmanship came into play. There is rivalry while the tournament lasts but little transfer to other aspects of the relationship.' The boys had had 'specific training in discrimination between situations of competition in sports (where aggression is socially approved) and other interpersonal or intergroup relations …'

Brian agreed that in baseball the concept of fair play was paramount. 'Once the game was over, you had to line up and shake the hand of every other person on the opposite team, no matter how bad you might have felt if you'd lost.' It was an inclusive and forgiving sport. 'I was never really good at baseball,' Brian told me, 'but I was never made to feel bad about it. People encouraged you for trying.'

Doug was insistent that the experimental team had made a major blunder. 'They missed the point. The point was to try to have us lose sight of what we'd been taught since we were little boys, and that was sportsmanship. You might want to fight tooth and nail over your ability to win a ballgame over another group, but you would not fight physically with them afterwards because that would have made you a lesser person. Your victory would have been snatched right away from you, you would have been considered a loser, not a winner. So I think whoever put two groups of kids and had them fighting over sports didn't understand American kids,' Doug told me emphatically. 'I know that sounds pretty romantic, a romanticised version of what took place, but the facts bore it out, didn't they?'

Later that same afternoon Harry Ness agreed to call a meeting for the boys to 'discuss their complaints' in the mess hall. Carper hid in the rafters to watch, and turned on the tape recorder they had secretly installed there.

At first, it was pandemonium. Listening to the tape, I had to take the headphones off as the voices roared in my ears. The boys were lined up on each side of the long table, shouting accusations. The Panthers accused the Pythons of cutting the flagpole rope because they were sore losers. The Pythons, insulted, yelled back in shrill and angry voices. 'We did not! You did it!'

'Why would we cut our own flag down?' one Panther yelled. I pictured Nathan, always quick to anger. 'We paid for it with our own money!'

'No one in our group has a knife!' a Python retorted, and their individual voices were lost in the roar of protests.

Peter, the oldest Panther, sat down abruptly at the table and tapped on it until the yelling died down. But the boys on both sides shifted and muttered. Peter looked around at the boys in his group and then waved his hand at the Pythons, who were lined up on the other side of the table. 'If any of these guys had cut it, they would have told us. They can't keep it to themselves.' He grinned at Laurence.

There was some laughter from both sides of the room.

'We're not such sore losers that we'd go and do something like cut the rope.' Laurence pulled out a chair and sat down too. Around him, the other boys muttered their agreement.

'Who said we did it?' Laurence wanted to know. 'Where did you get the idea we had a grudge?' He looked at Peter and the rest of the Panthers.

The Panthers looked around at one another but no one seemed to remember exactly where the idea had come from. If anyone remembered that Mr Mussee, the caretaker, had implied it was the Pythons, they didn't say.

Peter stood up abruptly and called his group into a huddle and they whispered intently, arms around one another's shoulders. The Pythons looked on, pretending they weren't trying to listen. The huddle broke up and Peter announced, 'Okay, if you all swear on the Bible, we'll believe you didn't cut the flagpole rope.'

One by one the Pythons stood and put their hand on the Bible, which rested on a folded American flag. Laurence went first, wriggling to get comfortable and clearing his throat as if he was about to make a speech, causing the boys to laugh. Then he said solemnly, 'I swear by the Bible and the flag that I did not cut the flagpole rope.' A wobbly cheer went up. John went next, drawing more applause. As each Python boy made his pledge, the others clapped and whistled. When it was Tony's turn, he added 'the Father, Son and Holy Ghost' to his pledge, and there was approving applause. Ill will between the two groups evaporated: Carper observed from his hiding place in the rafters that 'this reduced most of the violent hostility'. Any conflict had fizzled.

Carper's description of the meeting in the mess hall would offer no comfort to Sherif. The boys had resolved their differences and disappointed the researchers' expectations.

Sherif, desperate to 'increase hostility', told Ness to announce after supper that night that the staff 'forgot' to add some events to the tournament. They had craft activities and songs and skits already scheduled, but added more competitive activities. Ness told the groups that Capture the Flag was added, as well as a second tug of war and a treasure hunt. Then Sherif hurried away with Herb Kelman.

The Panthers were immediately suspicious when the additions to the tournament were announced. They guessed that Ness was favouring the Pythons, giving them a chance to get ahead. They were right.

The day before, after his baseball injury Doug had ended up in the infirmary a second time in as many days — which explained to me why he thought he had been in hospital. Once again, his homesickness had flared, and Sussman had convinced Sherif they were better off letting the boy go home, in case his homesickness became 'contagious'. Doug's parents were arriving that night to take him home. While Ness was making his announcement about additions to the tournament, Harvey was walking the 'downcast' Doug along the track towards the road. Harvey wrote that the two caretakers, Mr Mussee and Mr Herbie — Sherif and Kelman — caught up with them and offered to carry Doug's things. Doug seems to have been having second thoughts, especially given the Panthers' victory that day and the chance to be with his friends from the other group again. But it was too late; his father had likely already arrived and was waiting, engine running, at the turn-off to the camp.

Sherif, intent on gathering 'data' from the boy before he left, asked Doug why he thought the Panthers were winning everything, clearly hoping for an answer that showed 'glorification' of his group. But Doug shrugged. 'Luck, I guess … I guess we have more confidence.' Harvey picked up the questioning. What did Doug think of his Panther group? 'They're all good guys, every one of them, even in the other group, I love them all,' Doug answered.

Night was falling, and I imagined they walked the rest of the way in silence, the air fizzing with insects, Doug regretful about not getting a knife, and Sherif thinking it was best that the boy went home after all. When they saw the glow of headlights, they would have stopped, and Sherif and Kelman would have handed over Doug's bags.

Doug doesn't remember that evening, but he looked up Muzafer Sherif online and even though 'it seems impossible … but darned if I don't think I recall that face. He looked serious, dark, a little scary.' I imagined this memory was from Doug's last glimpse of Sherif, turning away with a scowl in the half dark and melting into the shadows, leaving Harvey to see Doug off.

The next day, Monday 2 August, the third day of the tournament was rigged in favour of the Pythons to 'increase morale'. Sussman and Sherif decided that keeping the scores of the two teams level would increase aggression and competitiveness between them. The Panthers were given a longer route in treasure hunt, the clues were harder to find, and staff deliberately slowed them down. Camp inspection was scheduled at the same time as the Panthers' kitchen duty, so they had less time to clean up their tent. By the time the ballgame was announced, the Panthers, who were already commenting on the favouritism Ness was showing to the Pythons, refused to play if Ness was umpire, accusing him of 'cheating them' and calling him 'a dirty bastard' among themselves.

Sherif engineered another frustration episode before dinner, taking items of the Panther clothing and hiding them in the Pythons' tent, then smearing the Panthers' table with the Pythons' leftover food. But the Panthers didn't take the bait. They attributed the missing clothes to a mix-up with the laundry, and while they were irritated about the mess on their table, and threatened to do the same to the Pythons, they never followed through on it.

On the barometer in the mess hall, the Pythons, until now the losing team, began to pull ahead. But the adults' tactics had become increasingly clear, particularly to the Panthers, who now expected that they would be discriminated against. I imagined that the Panthers would have been feeling angry, but also vulnerable, given they were far from home with adults who seemed biased against them. They were only ten and eleven, too young to have developed the teenage bravado that could hide feelings of fear or anxiety. Sherif seemed not to have noticed that he was testing a completely different group dynamic — the effect of discrimination by a group with more power and authority on the self-esteem and performance of a less powerful group.

The boys' unhappiness showed itself in their treatment of one another too. Kelman wrote that instead of directing their anger to their opponents, the Panthers had turned on one another as well as blaming the adults. This competition phase was supposed to bring each team together, but after each game there was recriminations and bickering.

There was bullying happening in Carper's Pythons group too. 'I had a fight with one of the boys, another Python,' Walt Burkhard told me. 'I don't remember what it was about, but I remember getting hold of a t-shirt of his and cutting it up with a pocket knife. I don't know what it was I was so angry about, but it was so out of character for me, especially pulling out a pocket knife to do some damage. I've never done anything like this again.'

It was hard to imagine Walt, this rather shy, softly spoken man, 'raising hell'. He had a youthful face, hatched round the eyes with fine lines, and bristly grey hair. Even though I got to the San Diego café before him, he was not the type to approach someone he'd never met. Instead, when he arrived he took a table outside and waited for me to come and find him. 'You can ask me any questions you want,' he said. But he paused over each snippet of memory or vague recollection as if weighing it for accuracy.

Walt looked troubled, recalling the self of this 1953 summer camp. As a professor of computer science, he was used to problems with clear solutions, but the behaviour of his eleven-year-old self in the summer of 1953 was a puzzle he was still trying to solve.

By the fourth day of the tournament, as the antagonism between the two groups failed to materialise, tension was building among the staff. On one side of the divide was Carper and Kelman; on the other was Sherif, Sussman, Harvey and White.

The archery contest that day was a turning point where the differences between the two groups of men became clear, with Kelman and Carper maintaining their role as observers, and the others acting more like participants. The clearing where the archery range was set up was small, a patch of short grass encircled by broad, leafy trees. It was cooler, with the wind stirring the trees, striping the edges of the grass with bands of shadow and light. Carper observed that there was a lot of 'fraternising' going on between boys on opposite teams: they helped one another retrieve arrows and yelled encouragement to their opponents. Kelman wrote that 'there was very little name-calling or serious hostility on the part of either group'. Perhaps because the archery conjured happy memories of the first couple of days, the boundaries between the two teams vanished. Boys on both sides cried out in admiration when an arrow landed with a 'puck' in the target, no matter which team the archer was on.

Jack White decided to take action. He tried to round up his Panther group, summoning Peter, who was coaching Laurence on how to better hold his bow, and telling Irving, who had joined Walt and Eric in the Python line, not to talk to the other team. But Carper and Kelman made no effort to stop the two teams from mixing. There might have been little hostility between the two groups of boys, but between the staff tensions were brewing.

During the fifth day of the tournament, amid a game of football, it got too much for Marvin Sussman, who likely saw the whole summer of hard work about to be wasted and his book chapter with Sherif slipping away. When he arrived partway through the game, Sussman was infuriated to see Carper allowing boys from both groups to crowd around 'in a concerned manner' when a boy was hurt during play.

In Sussman's eyes, Carper was clearly encouraging a friendly atmosphere and wasn't doing enough to promote rivalry and retaliation among the boys. Carper described the many 'gestures of friendship' between the two teams, but Sussman believed he should have been doing more to stop it. Sussman's notes for Sherif bristled with impatience and irritation: 'The game was being slowed down by the poor refereeing and the lack of enthusiasm of some of the counsellors in charge of the boys.'

Sussman took over, replacing Carper as referee and speeding up the game so the boys had to play faster and harder. He also overlooked rules in favour of the Pythons to keep the overall tournament scores neck and neck. Sussman wrote a particular note at the end of his description of the game, pointing out that if he, Sussman, had been in charge of the game from the start, it would have 'resulted in the hostility desired for this stage … and the conflict situation which was called for'. He went on, 'In terms of the experimental design this game should have been speeded up, the boys encouraged to "fight" their hardest, much as a good coach does with his football team, and the football rules should have been overlooked when necessary so as to satisfy the conditions set for this activity and stage.' In short, Sussman was saying to Sherif, let me take over. With me in charge, the experiment will be a success.

That night, Sherif gave his approval and told Sussman to replace Carper. Sherif's reason for getting rid of Carper was not because he wasn't intervening actively enough: he and Sussman, unable to countenance the idea that some boys might have seen through the ruse, accused Carper of spreading the 'rumor' that the camp was an experiment. 'The expose of the experiment became a "joke" with some of the members of staff,' Sussman wrote sourly. Years later, Sherif told a graduate student that the study failed because a 'fellow from Yale told the kids that they were being studied'. But Carper's notes were littered with examples of the boys' own suspicions about the camp, from the first day when Harold asked what the microphones in the rafters were for, to the accusations that they had been separated to see how they would respond, to the unfair penalties imposed by the adults in games. Sherif could not believe that the boys would have noticed anything without one of the adults giving the game away. On the other hand, Carper noticed how alert, curious, and observant the boys were to the adults and their surroundings. The day before, Carper had written that when Harvey approached some of the Python boys with the idea of a boxing match, one boy 'introduced the subject of trickery' and others chipped in with their suspicions, saying, 'You want to make us fight the others.'

Maybe it was the heat in the small, stuffy office where he and Sussman met that day after the game. Maybe it was that Carper felt sorry for Sussman and the impossible job he had taken on — he was likely looking exhausted after almost three weeks of sitting up until the small hours of the morning, night after night, with Sherif drinking and pacing and agonising, and going over and over every small detail. But Carper didn't protest when Sussman told him he was being replaced. Perhaps Carper was relieved to give up what by now he saw as a charade. He had little tolerance for duplicity. He had rebelled against the hypocrisy of his father, a pious and peace-loving member of the Mennonite community in public who, behind closed doors, beat his son often. Whenever he went home, he enjoyed showing off his secular lifestyle — smoking, offering his nephews and nieces sips of liquor from a flask. But Carper bit back his scepticism about the experiment that he would in fact describe in later years to colleagues as a 'joke', and agreed to take on whatever other duties were needed before the experiment ended.

As for Herb Kelman, he had been taking his role as 'scientific conscience' too seriously for Sherif's liking, especially when Kelman objected to the men's behaviour at the nightly staff meeting after the archery contest. 'I pointed out that they were going beyond observation of behaviour. You're supposed to observe, not to directly influence the way the boys respond because that's manipulation. There were points at which OJ and

Jack, who were students of Sherif's and very close to him, would push things along … to encourage certain kinds of behaviours … People don't necessarily do it consciously, I've great respect for these people, but they were students of Sherif's and dependent on him. They were in a very delicate, tricky situation.' Kelman didn't say it, but Sherif's watchful presence around the campsite as well as his own spontaneous interventions would have encouraged them too. But I was surprised when I read that Sherif, after this staff meeting, decided to exclude Kelman from all future meetings. Sherif had anticipated the potential for this kind of bias to happen at the start and had engaged Kelman to guard against it. Yet the very thing that Sherif had wanted to avoid he found impossible to resist.

As far as Kelman was concerned, he was just doing his job. 'My role was to point out, "Come on, you're not supposed to do that, it's like getting into the maze with the rat and pushing it. And you're not supposed to push the rat."'

But it seemed this was the last thing Sherif wanted to hear.

On the sixth day of the tournament, the activities were comparatively low-key — making model planes and performing songs and skits. Harvey, replacing Carper as participant observer, and White actively took on the role of coaches, cheerleaders, and combatants in charge of their teams, with Sussman's enthusiastic support.

It was the last day of the tournament. Sherif and Sussman had agreed the night before that the Panthers were at risk of giving up on the competition because they believed the men were favouring the other team, and as they had been susceptible to the 'contagion' of homesickness, they should be allowed to win. So they stacked the odds against the Pythons — the models the Pythons were given to assemble in the arts and craft contest were more intricate and difficult to finish by the deadline. In the mess hall, each team sat at a table on either side of the room, Harvey egging on the Pythons, 'congratulating and encouraging' them as they showed him their work, while White and Harvey traded 'loud' comments on the progress of each group.

At 2.00 pm, both groups gathered outside the rec hall for the announcement of the winners. When Ness read out the final score and announced the Panthers as winners, 'there were wild shouts and some of them jumped up and down,' Harvey wrote. As well as the 'beautiful jackknives', they were given a $10 cheque, which they ran and gave to Jack White for 'coaching us and helping us win'. They asked Ness for Doug's address so they could send him his knife and suggested to him that the Pythons should also receive a prize even though they hadn't won.

The losing Pythons were subdued. They sat around on the ground, pulling on blades of grass and not looking at one another. But the Panthers didn't gloat. Peter and Brian led their team over to shake the Pythons' hands and 'commended them for their good performance', praising their skills.

Sherif, observing from the back door of the mess hall, would have watched this with a rising sense of panic and anger. It was all wrong. Here they were at the end of the third stage, when the competitions were supposed to have got the boys riled up about their opponents. To the Pythons, the victory of the Panthers should have felt like salt rubbed into their wounds. Anger was supposed to spill into violence, fighting, and retaliation. Sherif had promised this much, and more — a dramatic denouement where he was able to bring together fractious and warring groups into a harmonious whole. In the final stage, when the hostility and hatred between the boys had reached fever pitch, Sherif planned to set fire to the forest so the boys would be forced to cooperate to save their campsite and join forces to put out the flames.

In his efforts to get one group fighting the other, Sherif and his staff had cut the rope on a precious flag, 'stolen' items of laundry, and smeared and demolished the food set up for one group's dinner, all to trigger a fight between them. Each time, the boys' irritation and anger was transitory. The only enduring resentment among the losing Pythons was against the staff. Now the Pythons were complaining they had been treated unfairly. Instead of blaming the Panthers, they attributed their loss to the actions of the adults — in particular, the unfair rules imposed by Ness during the tug of war; the favouritism of Sussman, who gave them noticeably more difficult models to assemble in arts and crafts; and the treachery of Mr Mussee, who, being a caretaker, would naturally be short of money and open to bribes from Sussman to vote against them in songs and skits.

The irony was that the boys' identification of when the adults were rigging the results was — apart from the bribery of the caretaker theory — spot on. They'd been treated unjustly and they knew it.

So what happened next should have come as no surprise.

OJ Harvey sat around the campfire on the hill after supper with the dejected Pythons. It was a warm night, but he built up the fire and sparks flew skywards, the light casting an orange glow over the boys' glum faces. Harvey was trying to encourage someone to tell a ghost story, but the boys weren't enthusiastic. A sudden burst of noise made them all jump. Irving, one of the Panthers, burst from the trees, yelling and sobbing. 'You'd better get down there!' His glasses flashed orange in the firelight. 'Or we'll wreck your tent!'

The boys around the fire jumped to their feet and began running down the hill, racing towards their campsite. Harvey hurried behind them. Down in the staff office, Kelman heard the shouting too and hurried outside to see what was going on. Kelman and Carper, now both outsiders, had no knowledge of the night's plans.

Laurence, Walt, Tony, and the other Pythons raced to their tent. The Panthers were there waiting for them, shouting and jeering. The Python tent gaped open, the pale shapes of their bedding and clothes strewn about in the dirt. It was impossible at first to tell what was going on. Boys on both sides were shouting and crying out their dismay.

'Are you crazy?' Laurence cried.

'Now you've done it!' someone yelled.

'What happened?' Mickey stood with his mouth open.

I imagined the boys ranged on either side of the tent, their eyes glittering, the din of their raised voices, their pounding hearts. And Sherif in the shadows, holding his breath.

Amid the shouting, Laurence and the rest of the Pythons demanded to know why the Panthers had wrecked their tent. Anger distorted the features of the other boys, who faced them so defiantly. Over the noise, some of the boys pieced together the story. The Panthers had arrived back at camp after their celebratory marshmallow roast and found their own tent demolished and their belongings strewn about. They assumed the Pythons had done it, so they ran straight to the Pythons' tent and retaliated. Laurence, whose lower lip was trembling, yelled over the top of their cursing and crying, 'But we were up there! All of us.' He pointed up the hill, to where the fire glowed among the trees.

The boys went quiet. Fear flickered from one boy's face to another.

'We were scared,' Brian Wood told me. 'If it wasn't one of us, then who was it?' They were no longer Pythons or Panthers; they were a group of children in the dark at the edge of the woods, where shadows seemed to be gathering. A shiver ran through the group and the boys instinctively stood closer together.

'Come see,' Peter said into the quiet, gesturing back towards the tent on the other side of the stream. 'Come see!'

Then the boys were running again, but this time in a single throng, their shouts echoing through the woods as they poured across to the other side of the camp.

At the Panthers' tent, it was pandemonium. The boys cried out in dismay, 'See? See?' The tent was flattened, suitcases had been thrown among the bushes, bedding was strewn around in the dirt. Irving picked up his broken ukulele and began to cry. The boys stood in a huddle, their faces pale. Peter turned to Laurence and the rest of the Pythons and said shrilly, 'Do you see now why we're mad at you?' Laurence said that he could see it.

The boys spread out in twos and threes and began picking up clothing and shifting beds. Some moved around the base of the tent, preparing to lift the centre pole and get it standing again.

One boy went to fetch a lantern. All of them moved around the tent gingerly. Who could have done this? they asked one another.

Harvey arrived and called out that the Pythons should return to their own area, but the boys didn't reply. White was doing his best to keep the groups separate too: he announced that the Panthers could look after it themselves. Harvey persuaded the Pythons that they should get back and sort out their own tent, but they left reluctantly, promising the Panthers they'd be back as soon as they were done.

Kelman stayed behind at the Panther tent, helping to set it straight and eavesdropping as the boys discussed who could have done it. Peter and Irving shook the dirt out of the sleeping bags. They had decided the other group had nothing to do with it. Peter said, 'Laurence said the Pythons didn't do it and he doesn't lie. Laurence is no liar.' Irving agreed. White, who was smoothing out trampled pillows, said no one else could have done it, but the boys yelled back at him: 'They couldn't have done it, because if they had they wouldn't have offered their help!'

Then Kelman went to the other side of the stream, to the Python tent, to help the boys tidy the mess. They were agitated. Eric, whose accordion was untouched, was upset for Irving and his broken ukulele, and said angrily that he would go around and demand an alibi from all the adults in the camp. As they retrieved things from the ground, they kept up a staccato discussion, considering and just as quickly discarding different theories. Perhaps it was Sandy the cook, someone said, who had told the boys he was angry that he hadn't been able to go to the races at Saratoga Springs. Someone else suggested the caretakers. Eric said aggressively to Herb Kelman, 'Where were you?' Laurence turned to Harvey and asked whether this was an experimental camp. 'Maybe you just wanted to see what our reactions would be.' I imagine Carper would have busied himself moving the lantern, or straightening a suitcase, not trusting himself to answer. When the boys told Carper soon afterwards that they were going to help the other group put up their tent, he made no move to stop them. Meanwhile, Harvey went in search of Sherif to tell him he hadn't been able to stop the two groups from helping each other, bumping into Kelman on the way. Kelman told Harvey that the Pythons should not be stopped if they wanted to help the Panthers straighten up their tent. Harvey included his curt reply in his notes of the evening: 'I replied that if I stopped them, I would specify such in my report', before he hurried on.

Harvey heard Sherif's voice, high and rising, before he rounded the corner of the mess hall and saw the men silhouetted against the light from the windows. Sherif had clearly heard the news. Sussman was saying something in an urgent voice, but Harvey couldn't catch it.

'Dr Sherif,' Harvey hissed, hoping Sherif would lower his voice. But Sherif was oblivious. Harvey's mouth went dry when he saw Sherif's face, red with fury. 'A vulture,' Sherif spat the words out.

Sussman's face was pale and shiny with sweat, and he kept pushing his glasses up his nose.

'A greedy vulture!' In Sherif's mind, Sussman had ruined the experiment out of his selfish lust for credit. Everything about Sussman, from his youth to his ambition, now infuriated Sherif. He took another step towards Sussman, shaking both his fists in the air around Sussman's ears.

Sussman nervously licked his lips and took a step back. 'But we agreed!' Sussman protested. 'Dr Sherif, we agreed before supper —'

'I never said!'

'Dr Sherif,' Harvey said more loudly, skirting the fire ring and wood pile.

Sussman looked imploring at Harvey. 'OJ, tell him.'

But Harvey pursed his lips and gave his head a little shake as if to say, it's no use. Sherif had been drinking, and Harvey knew that trying to reason with him in this state was hopeless. He'd seen Sherif furious and on the attack before, but never like this. This night, Harvey told me, Sherif had gone 'bonkers'.

Sherif drew back his fist, ready to take a swing, and Harvey picked up a piece of wood from the pile and grabbed Sherif's arm. 'Dr Sherif!' he said, pulling on his arm to get his attention. 'If you do it, I'm gonna hit you.'

Sherif tried to shake him off, but Harvey held firm.

'We discussed it ...' Sussman was babbling. 'We did ...'

And they had, just an hour earlier. Harvey wrote in his notes:

> Around 7.00 pm, before Sussman talked to Sherif, I met with Sussman by the mess hall and we discussed the frustration planned for the night. The consensus had been reached between Sherif, Sussman, myself, and Jack White that the frustration should be against the Panthers since they had won the tournament and seemed quicker to get angry and seek revenge against the other group.

While the boys were engaged at their marshmallow roast, Harvey wrote, 'Sussman was to go to the Panthers' tent and wreck it.'

The competition and now the wrecked tent had not only provoked no hostility, but the boys were reconciled, and had turned on the staff. I thought of Kelman's remark about how Sherif was unable to accept that he might have been wrong. Even if the tent-pulling incident wasn't his idea, he'd agreed to it, but now he looked desperately for a scapegoat. And Sussman was it.

Sherif abruptly stopped struggling against Harvey and then turned and rushed away. At 11.00 pm that night, he called all the men into the mess hall and announced the experiment was over. OJ Harvey, Jack White, and even Marvin Sussman argued that they should continue, since they'd come this far, but Sherif, despondent and looking exhausted, refused to listen.

I thought of the letters Sherif had written in the months leading up to the experiment, his mood of foreboding at the prospect of what he called a 'Herculean' task. I had thought, reading these letters, that this vision of himself as the lonely hero labouring against the odds and his at times extravagant and sentimental language had seemed overly melodramatic. From his letters and writings, Muzafer Sherif saw himself in dramatic terms, as a social scientist labouring to reveal a profound truth, someone whose work was his life. This study was the end product of three years of funding and countless years of work. The failure of this study was more than a setback; it would have felt catastrophic. Still, he first blamed Carper and Kelman, and then his right-hand man Marvin Sussman, for the study's lack of success, unable to admit even in the face of all the evidence that the theory he put forward was the element at fault.

'He was devastated,' OJ Harvey told me. 'He hardly spoke for the next two days. He took it very badly and was deeply depressed.' Sherif stayed in the tent by the edge of the birch grove and made it clear he wanted to be alone.

Meanwhile, each of the men used the remaining days of the camp to deal with what had happened in their own way. Harvey and White grappled with what they would do now for their dissertation. Sussman did his best to keep himself busy and useful, trying to ignore the fact that Sherif was blaming him for the study's failure. Kelman and Carper wrote detailed notes summarising how and why the study failed so that Sherif had an account to guide him in case there was a next time.

In the last days of the camp, the boys felt abandoned. Jim Carper and Jack White no longer showed any interest in them, and looked tense and unhappy. Even the caretaker, who had been everywhere they went during the first couple of weeks, was nowhere to be seen. Walt remembers roaming aimlessly around the camp with a bunch of other boys, and at one point stopping to throw bricks at the old upright piano near the mess hall. 'Why we did this I don't know, but we destroyed that piano.' Walt, the son of a piano teacher, who had been learning the instrument for five years before this camp, sounded mournful. I imagined the sound of the bricks hitting the keys, the crash of discordant notes, the piano booming and shrieking, the lid caving in, the front board gaping, the hammers splintered into pieces. Walt is still unsettled by the memory of it: 'It was so out of character for me.' I imagined how loud the silence would have seemed when it was over, with the boys looking at the piano split and tumbled in the dirt as if they were waiting for something to happen, but it would have been quiet, just the tick of insects and the rustle of the leaves nudged by the hot wind. 'There was a sense we had of being on our own. There was no one stepping in and saying, "Hey, you can't do that."'

And it was that moment that was most unsettling — that feeling that nothing they did mattered, no one would stop them, no adult was going to intervene and tell them what was or wasn't against any rules. 'When it was over I was not happy with myself. It was so out of character,' Walt repeated.

But there was a sense of freedom too. Brian remembered the relief of all being back together as a single group, and wondered if it brought them closer together. 'It reminds me of that sense of kinship you read about, when strangers have been through an experience together and they develop a bond. When it came down to it, we stuck together, didn't we?'

Brian's words stayed with me. Yes, the boys had stuck together, but the same couldn't be said for the men.

For Sherif, remaining at the campsite was an unbearable prospect. The next day he wrote in his diary: 'Talked with OJ and Jack in Panther campfire area from 2.00–5.30.' They had sat on a log facing the cold fireplace. The day was overcast and the wind shivered the trees. Sherif sat with his elbows on his knees, smoking and staring into the fireplace. Harvey did most of the talking. Sherif didn't take much persuading. 'He was absolutely heartbroken,' Harvey recalls. 'But we all agreed we wanted to do it again. Sherif agreed that he had been pushing too hard and pushing all of us too hard and that he wasn't the best one to be in charge.' They went back and forth, 'salving their wounds', going over and over what had gone wrong. And what had gone right. 'Jack and I told him we'd only do it again if I could be in charge.'

Sherif left the camp later that afternoon, leaving Harvey and White to dismantle what was left of the study.

When Doug heard the story of how the camp ended, he was thrilled. 'I would have loved to have been part of that. I feel like I was there, talking about it. I've thought about it a lot the last few months: how it was just a few years after World War II, and atrocities and what people will do to others was not far from people's minds, but to know that the boys had absorbed the sportsmanship concept — I think that's what it is, anyway — to a greater extent than the combative concept is really appealing to me. It's been a great opportunity to rethink a part of my life. It's led me to think about a lot of things in different ways, good ways. It's very uplifting!'

We were sitting on the couch, facing the window that looked over his lush summer garden. Doug pulled at the ears of his little black dog, resting at his feet. But what exactly his parents knew about the camp still bothers him. 'I once described my memory of the camp as a "dark memory": not sharp and vivid but murky, unpleasant.' Doug paused and scratched the dog under the chin. 'And no one in my family ever talked about it afterwards. It was not made part of our family history, which is very unusual — except for my dad's experiences in World War II, our family talked about everything. But not this camp. It's like my family blocked it as much as I blocked it.'

Doug, like all of the boys I'd spoken to, didn't remember his parents telling him that it was anything other than a summer camp. But Doug thinks his father likely guessed. It's a theory he came up with after talking it over with his brother and sister. 'My father revered doctors his entire adult life. He was in the medical field — he was a pharmacist — and all his friends were doctors. But he considered psychiatrists and psychologist to be "quacks", people who didn't know what they were talking about, who were harmful and of no value. And I think it's very possible that once he found out that he got duped into sending me off to this thing, he held that against the profession.'

But Brian believes his parents never knew, and he wondered about his own role in keeping them in the dark. 'They would have asked me when I got home what it was like, and I wasn't the kind of kid who kept secrets — not at that point, anyway.' Perhaps he felt protective, I suggested, he didn't want to make them feel guilty? But it was as if I hadn't spoken. 'I could talk to my mother about anything,' Brian said. 'Why didn't I talk to her about this?'

Sherif set out to disprove theories that prejudice and conflict sprang from human nature, an authoritarian disposition, or a pathological personality, or were an expression of displaced frustration. Prejudice, hostility, and violence were the result of the attitudes and relationships between groups of people in society. Take a normal group of well-adjusted people who are friends, he argued, put them in groups competing against one another for a valued prize, and hostility, hatred, and even violence is inevitable. Friends will become enemies. For Sherif, the experiment was a failure because it did not deliver a scenario he had clear in his mind. In Sherif's scheme of things, a 'crucial' test of his theory was that the 'budding friendships' between the boys in the first days of camp would be reversed.

But none of the men seemed to have noticed that the staff enacted their own scenario of group prejudice and conflict. Initially a cohesive team focused on the goal of a groundbreaking study, under the powerful personality of Sherif the staff group fractured, Kelman and Carper were shunned and sidelined, and Sussman became a scapegoat for Sherif's frustration. The only violence in the camp was Sherif aiming to slug Sussman and Harvey threatening him with a block of wood.

I lost touch with Doug for a while after my visit. When I heard from him again he told me he'd been out of action, first in hospital, then for months in rehab. He'd been racing his mountain bike up a climb in the western part of New York State, ignoring the pain in his chest, until he collapsed with a heart attack. By the time they got him to the little local hospital, the staff decided he was 'gone', but a visiting cardio specialist injected him with 'clot busters' and saved him. He'd had a quintuple bypass requiring nine hours of open-heart surgery. After four months of rehab, he was just about to go back to work. He was playing it down, but it was clear that Doug had almost died.

'You know, I don't know if learning about this experiment had anything to do with it,' Doug said in passing. Doug has a habit of saying serious things in a joking way that makes it hard to know whether to take him seriously or not, and by the time I'd drawn breath to follow this up, he'd moved on again. They say after open-heart surgery that all that handling and prodding of your heart makes you feel depressed. But was this what I had done too, prodding and poking around, getting into people's memories, dredging up half-formed recollections they would have preferred to let lie?

Doug said that being off work and recuperating, he'd had a lot of time to think, and he'd come to a different conclusion about the experiment this time. 'It was wrong, by which I mean not just poorly planned but poorly judged. They misunderstood human nature. They certainly misunderstood children. Ultimately they were able to manipulate children, but they were morally wrong. And that's where I came down on it.'

So his feelings about it had changed?

'Yes, and one of the things that changed it was having gone through my own health situation. I've learned from talking to somebody who gave me some great advice recently about this, that if you've had something significant happen to you — in this case it was trauma — the whole bit about going back through it cathartically is a mistake. All it does is make you relive it. It never diminishes it, it enhances it. It's better to respect it, understand what happened, and then get on with your life because you didn't die and you're okay. Keep moving. And I think that translates to this camp thing for me a little bit. So while I was originally fascinated and glad, it doesn't feel the same, it feels more like I was used. Not abused, but used. And that really makes me mad.'

I was struck by the shift in Doug's mood, the sense of anger and loss. I felt responsible in the sense that here I was, years later, unwittingly caught up and catching Doug up too in the moral ambiguity of Sherif's experiment and its legacy. It didn't matter how much I tried to be objective, I was tangled in the net of this history and its consequences.

I thought of the other boys at this summer camp, who had

bonded in the face of the men's manipulation and defied the efforts of Sherif and the others to make them betray their friends and turn on them as enemies. Initially I construed that as a kind of victory, a triumph of the power of friendship to override adversity. A good-news story.

But the fact that the boys had resisted did not mean they weren't affected by the experience. Perhaps they had all put the summer of 1953 out of their minds, and this explained why for the boys I had spoken to, it was an effort to recall. It hinted not just at the ravages of time and memory but also at a suppressing of something they preferred to forget. Their participation in an experiment they had been unable to truly consent to had come at a cost. Debriefing the boys had been out of the question, given that Sherif had decided to run the study again, so they were sent home carrying doubts, secrets, unanswered questions. And now my presence — the researcher, a character in the broader frame of this story — had prompted an unofficial reckoning for some of them.

How many other 'lost boys' did this era of social psychological experimentation generate, I wondered. How many psychological wounds were caused in the pursuit of scientific and historical understanding?

# 7
# The Robbers Cave

A year after Muzafer Sherif cancelled the Middle Grove experiment, he was behind the wheel of a university station wagon, heading south-east from Oklahoma City to Robbers Cave State Park. Sherif planted his foot on the gas. He drove like he was racing towards the future, the past disappearing in a cloud of dust behind him.

'He just wanted to forget it,' Herb Kelman says, leaning back in a chair beside a squeaking air conditioner that was losing the battle to keep the lounge room cool. The caretaker in the 1953 photos at Camp Talualac now has white hair swept back from a bald pate. He is clean-shaven, but he wears the same kind of heavy-framed black glasses that he wore back then.

The details of the last dramatic last days of the Middle Grove experiment are not in any of Sherif's files and boxes in the archives at Akron. But Herb Kelman has held onto his notes for sixty years because he was convinced that Sherif should have written about the study instead of trying to bury it. He's dug out his notes and papers in preparation for my visit, and jokes that I've reinforced his habit of never throwing anything away. These days, Kelman has an international reputation as a pioneer in peace research. Since he retired from Harvard, he's been trying to go through and sort all the papers and files that he's brought home with him to his Cambridge apartment. At eighty-five, he is just back from Vienna, where he was awarded a Gold Medal of Honor from the city for his work in international peace research.

'My feeling was an interesting thing happened, a terribly interesting thing happened!' Kelman's voice rises with excitement. 'And it would be interesting to find out why it happened — these kids found their own way to reconciliation, and for anyone interested in these processes, it was a very exciting event! It didn't have anything to do with the original experiment, but it was a great learning opportunity.' He shook his head. 'Yet it went against Sherif's hypothesis and so he treated it as a failure. As far as he was concerned, the best thing to do with this study was to try and forget it. And to try as much as possible to blame it on others. It was easier for him to say it was a manipulation failure rather than to say, "There is some other variable operating here that I haven't recognised in my theory." Muzafer's reaction was, "I wish this hadn't happened. Take it away!"'

Like Herb Kelman, Marvin Sussman didn't want to forget the experiment either. He wrote to Sherif saying that even though he expected Sherif's 'continued mistrust' and blame for the experiment's failure, he believed it was important to write about its flaws: 'You would then be actually making another contribution to Social Psychology. It seems to me and others that advances in group research are made both with successes and failures ... Many social scientists have asked about the '53 study and expect some kind of report.'

But Sherif would have none of it. He was so intent on pushing the failed Middle Grove experiment out of his mind that six months after it was over, he still had not told the Rockefeller Foundation that he had aborted the study. It was only by chance that they found out. The Foundation received a letter of complaint from a man who had read about Sherif's study in news reports of the Foundation's 1953 annual report. The man wrote, asking:

> What is the professor's background?
>
> Were the children in the camp underprivileged or orphans? If they had parents, did the parents know what the professor was doing to the minds and characters of their children?
>
> What benefit to humanity does your foundation expect to derive from the expenditure of $38,000 on this study?

The Foundation contacted Sherif, asking for more detail about the experiment in order to draft a reply. In their exchange of letters, program officer Leland DeVinney was clearly dismayed at Sherif's answer: 'Do I understand correctly from your letter that the work last summer went only to the end of stage 2? If this is in fact true, is my inference justified that this must represent something of a disaster with respect to the main objectives of the grant?'

Sherif replied saying that he just needed a little more time to run an additional small study, and to both reassure and distract the Foundation, he included the manuscript of his soon-to-be published book 'Groups in Harmony and Tension', which he described as a 'manual' for researchers containing his theory about group behaviour that he was about to test in his upcoming experiment. He didn't mention that the major funded experiment — the 1953 study — was to be little more than a footnote in the book. Whether DeVinney bought this or remained sceptical, he reported to his superiors that

> Professor Sherif reported that a great deal had been accomplished to date as is evidence in the completion of the well-received book, Groups in Harmony and Tension, and in a number of research monographs and reports which are appearing in various journals. He feels it important, however, to re-check certain aspects of the field experiment during the coming summer before completing the project with the preparation of the final report …

Sherif might have temporarily appeased the Rockefeller Foundation, but this 1954 trip to Robbers Cave was a last-ditch attempt, with just the crumbs of the original grant money, to rescue the project. No matter what undertakings he'd made to Harvey to hand over the reins on this next experiment, his instinct was to tighten his grip, not loosen it.

'He was under enormous pressure,' OJ Harvey told me. 'He knew he had to make it work.' And OJ, as I came to know him, felt the same. 'We were both zealots, anti-Nazi. And we would have thrown ourselves off a cliff if it hadn't worked.'

OJ's deteriorating eyesight made him a reluctant driver, so he had arranged for his friend and former student Gerry to pick me up at my motel on the highway. A bearish white-haired man with a gentle manner, Gerry told me on the drive that he hadn't realised quite how much my trip meant to OJ until that morning, when his phone rang at dawn. It was OJ, checking once again that Gerry had the details right for my motel and pick-up time.

We drove along the highway towards Boulder but then took a right-hand turn along a country road, past white picket fences and occasional long, low houses surrounded by paddocks, set back from the road. OJ came out to meet us. I recognised him from the photo I had seen of him posing with Sherif and White. He had the same open and curious smile, as if he was expecting the punchline of a joke. In his eighties, he was still sprightly, and although his features were softened with age, I glimpsed the eager young man with his hands on his hips I'd seen in the photos from Middle Grove in 1953, looking determined and gazing into the distance, as if he could see a better future there. Despite fifty years in Boulder, OJ's Oklahoma accent is unmistakeable.

The mountains visible from his back porch were capped with snow. OJ had been a shrewd businessman as well as a professor of psychology at the University of Colorado, and the house was large and comfortable, with long leather couches, slate floors, a throw rug over a rocking chair. It was a far cry from the sharecropper house where he grew up in south-east Oklahoma. OJ was proud of his roots. He picked up a long stick with a string net at one end from where it leaned against the wall. It was a ball stick, he said, used in a traditional Choctaw game. He swooped it through the air. Did I know that Native Americans invented lacrosse?

But beneath his courteous manner, OJ seemed a bit nervous. I asked him what 'OJ' stood for, and he said his parents believed that if you named your child after someone bad, you inoculated them from the same behaviour. OJ was called after a notorious cattle rustler who preyed on poor families like his own, stealing away their livelihoods in the middle of the night.

Since then, I've discovered that OJ was the first Native American man to get a psychology PhD. His friend Jack White, from the Kiowa tribe, was probably the second. Not that OJ would have told me that fact himself. He has a horror of bragging: he says that's an 'Okie' thing. As a boy, he used to read to his parents at night because neither of them could read or write. OJ's father impressed upon him that just because he could read, it didn't make him any better than anyone else. How ironic that he teamed up with a professor who boasted that his theory eclipsed all other explanations of group behaviour, who dismissed Freudian psychological theory as 'one-sided', a 'failure' that lead to 'blind alleys'. Was this another reason Sherif and OJ worked so well together, I wondered, with OJ as a kind of ballast to Sherif's grandiose claims?

Despite OJ's protestations that he had forgotten a lot of important detail, his recall was impressive. We began talking about Robbers Cave, but first I wanted to know about the earlier 1953 study. The story he wanted to tell was the scientific narrative. But it was impossible to talk about the experiments and the ideas they were exploring without also talking about Sherif's personality. 'I'm getting personal now,' OJ would say, as if what he was about to confess was irrelevant. And I would straighten up, pay closer attention.

By the summer of 1954, Sherif's dejection and depression was gone. He was excited as he drove into what he called the Wild

West, towards that corner of the country close to where the borders of Oklahoma, Arkansas, and Texas cross. He had pushed last year's failed experiment to the back of his mind and was full of a nervous energy.

OJ sat in the passenger seat. In the intervening year, he had found an alternative experiment for his dissertation, a comparison of group solidarity among lower- and upper-class young men in Oklahoma; completed his PhD; and had a postdoctoral researcher position lined up at Yale for the fall. He stayed quiet and let Sherif talk. Sherif's confidence seemed to have returned. This time, he told OJ, things would be different.

Sherif trusted OJ, and asked his former student to call him Muzafer. But OJ was wary. He'd worked with him long enough to observe the bouts of paranoia and erratic behaviour that buffeted Sherif and then those around him. 'I had to keep him at arm's length because he was just so temperamental,' he confessed.

Sherif talked rapidly as he drove, waving his hands about and seeming to pay little attention to the road, telling OJ what they must do when they arrived. But OJ had his own plans for what would happen when they got there. He'd spent the past twelve months mapping them out. He was worried. It seemed that Sherif might have forgotten about their bargain.

Up ahead, OJ caught a movement. A mob of cattle was travelling slowly across a field and were headed for the road.

'Dr Sherif, there's some cattle up there.'

'I see them.'

'You might want to slow down.'

'Oh', Sherif said, 'I'm okay,' and kept his foot to the gas.

'Slow down!' OJ said urgently, as the steers moved into the

centre of the road. But Sherif didn't pause. OJ leaned across and slammed his hand on the horn. The steers scattered to the edge, but a big Hereford bull had stopped in the middle of the road.

OJ squeezed his eyes shut at the last minute and felt the car swerve. His body was flung hard against the door, then away again, as Sherif zigzagged up the road. When he opened his eyes, Sherif was thumping the wheel, exhilarated.

'Pull over, Dr Sherif,' OJ said through gritted teeth. 'Pull over.'

'I handled that well, didn't I?' Sherif said proudly, slowing to a stop.

But OJ didn't answer. He flung open the car door and stood out on the road, taking deep breaths until he felt calm enough to get back in the car and take the wheel. It seemed a metaphor for what happened when Sherif was allowed to take charge, and OJ vowed as he opened the driver's side door and slid behind the wheel that he wouldn't make the same mistake again.

At schools across Oklahoma City in the spring of 1953, OJ stood at the edge of the yard watching fifth-graders at play. This time he was looking for a different kind of boy than the ones they had recruited in the past. This time he would target athletes, boys who thrived on competition, with average grades. They would be 'normal' white Protestant and middle-class boys with no obvious physical or emotional problems; this time there'd be no bedwetters, no loners, no truants. Among the flurry of activity on the playground — the pock of the ball on the bat, the cries of the children — OJ watched to see who ran fastest, who rallied teammates with shouts of encouragement during the games.

Sixty years later, OJ was still proud of how he talked his way into the Oklahoma City school system and how he convinced the Director of Elementary Schools to write to principals in schools across the city, giving him access to detailed student records so he could select just the right boys for what he told them was research on leadership. Once he'd chosen a boy, OJ contacted parents by phone, an informal and more effective method than writing a letter in a place like Oklahoma City. The fact that he was from the University of Oklahoma opened doors. It was home to the Sooners football team and the legendary coach Bud Wilkinson, whose recent winning streak — including the first of what would be three national championships — had fanned a religious-style football fervour across Oklahoma, reinstating badly dented state pride. The Sooners were revered, and their home university basked in reflected goodwill. Add to that OJ's gentleness, his respectful manner, and his Oklahoma accent, and he won parents' trust.

OJ told parents 'the truth but not the whole truth', describing the camp as a chance to study which boys would become leaders and which would become followers. This might have appealed to parents' Cold War era anxieties, and all of them, OJ said, were pleased and enthusiastic about their boys taking part. He asked them to promise not to come visit because it caused homesickness. Welcoming a native Oklahoman into their homes, one who had none of the airs and graces they would normally expect of a university man, these lower-middle-class and working-class parents would have had no inkling that OJ was worried about the experiment ahead.

This time there was no nurse, no camp director, no Herb Kelman to play scientific conscience, and a much-diminished staff team. Things were both simpler and more complicated. Sherif and OJ had streamlined the theory they were testing to make it more achievable. They abandoned the hypothesis Sherif had started out with in 1953, that group loyalty overrides friendship. At Robbers Cave, they left prior friendships out of it altogether: this time, they wouldn't allow the boys even to meet before the competitions, let alone become friendly. But finding boys who didn't know one another — even if OJ did select them from twenty-four different schools — was a challenge in Oklahoma City in the pre-TV era, when the locale had a strong social network, and OJ ended up with just twenty-two boys.

Running the experiment with only four staff, OJ knew that a 'hands off' approach would be impossible. Participant observers Jack White and Sherif's newest graduate student Bob Hood — a former pharmacist from the small town of Guthrie who was a private, sweet-tempered man — would have to pitch in with the junior counsellors to run activities with the boys. Even with the best organisation, the day-to-day workload was going to be gruelling. Sherif's role, too, was more complicated. OJ worried that the boys would be apprehensive about someone with such a heavy foreign accent in a time when fear and suspicion towards foreigners was intense. So he introduced Sherif as his friend, who would help out with caretaking. But above all there was the pressure: this last attempt with the dregs of the money from the Rockefeller Foundation, and Sherif's reputation at stake. OJ was conscious that he was both the glue and the buffer that would hold the group of four together. 'It was an ingroup amongst us really, it was a case of high loyalty to Sherif. Jack and Bob weren't as close to him as I was, but they respected him and they were gung-ho.' Yet OJ knew that the ties that bound them could be tested.

Bill Snipes, a round-faced boy with a cheeky smile, sat in the back seat of the bus between a scowling boy called Red and a small boy with a toothpick that he worked in the corner of his mouth, called Hollis. There were half a dozen boys on the school bus that OJ hired by the time they got to north side of Oklahoma City. Unlike the boys from the 1953 study, many on this bus wore well-tended hand-me-downs and carefully patched jeans.

Northside was the wealthier part of town, and Red made a fuss when a group of northside boys boarded, shouting, 'Only southsiders up back!' None of the newcomers objected. Red was bigger than all of them and looked ready for a fight, so they meekly took their places behind the driver.

By Holdenville, Bill Snipes had left the back seat and was swaying in the aisle between a group of boys up front, leaning to look out the window every so often to announce the number of miles to McAlester each time he saw a sign. For Bill, the diminishing miles to their destination was something to celebrate. His parents had never been able to afford to send him away to camp, and he was excited to have been picked. For Smut Smith, a pale lick of a boy who had found it hard to say goodbye to his parents that morning, it was only the beginning of the realisation of exactly how far away this camp was.

A cheer went up at the end of the four-hour drive when the bus slowed off the dirt road and passed through the park's entrance, marked by a pyramid of logs painted white to spell out Robbers Cave State Park.

Before it was named Oklahoma, this was known as 'Indian' territory, the final destination on what Native Americans called the Trail of Tears, a dangerous and anguished journey for the hundreds of thousands of Native Americans forcibly removed from their ancestral lands. Pressured by white settlers who coveted the rich farmland, the tribes were forced by the federal government to leave their homelands and walk more than a thousand miles with few supplies to the new 'Indian' territory across the Mississippi. The Choctaw tribe established the Choctaw Nation here; Robbers Cave State Park was on their land.

The first time I visited Robbers Cave State Park, it struck me that it had a long history of intergroup conflict, especially after what had been thought of as barren new land turned out to be rich with oil and gold. I wished I'd thought to ask OJ whether and how the subject matter of Sherif's theory about animosity and prejudice between groups fighting over resources related to OJ's own life, and the poverty and dispossession of Native Americans he saw growing up in Corinne, not far from Robbers Cave, as a child.

The new territory had a reputation for lawlessness. Rich in gold and oil, it was a magnet for rogues. The San Bois Mountains, with their caves, became a hiding place for train robbers, bank robbers, gangs, gunfighters, and desperadoes. Just a few years earlier, the park's caretaker had found bank robber Pretty Boy Floyd's haul of stolen gold jewellery buried in mud when he was netting minnows in the park's stream. And the place was built by lawbreakers. It was carved out of the forest back in 1931 by a construction gang of fifteen prisoners from nearby Oklahoma State Penitentiary. For the prisoners, four white and eleven black men, who worked for months without a guard building the stone huts, digging and laying water pipes, and constructing fences, it must have been a welcome but brief taste of freedom before they were returned to the darkness of their cells.

The Division of State Parks brochure made the location's dark past sound romantic: 'The Robbers Cave ... is the legendary hideout of Belle Starr, the James Brothers, and other colorful early-day outlaws. Legends of hidden treasure, outlaw trails and gun battles with law enforcement officers stir the imagination of young park visitors.' I couldn't avoid the thought that the experiment that unfolded there was a continuation of this place's past, where the line between the good guys and the bad was blurry, and it was difficult to know who to trust.

Off the bus, the boys followed their counsellor between the trees to a stone cabin with a large chimney and a brown-shingled roof, which sat halfway between a recreation hall and a mess hall. Back then, the site was used as a boy-scout camp. Built of the same stone, with the same brown roofs, the camp buildings merged into the environment as if they were camouflaged. The air in the bunkhouse was dusty and close, and the boys would have headed off enthusiastically to the swimming hole. So too did Sherif, as he moved silently from tree to tree, tracking the boys as they made their way to the creek.

'We wouldn't call them a group yet. At first they were just a bunch of boys,' OJ said. 'So we had to do something about that.' At lunchtime, OJ and Sherif joined the boys on the pretext of bringing food and watched as they decided how to divide up the slab of ham, the unsliced bread, and the whole watermelon. 'It was deliberate,' OJ said. 'We created situations of interdependence where they had to rely on one another, to divide food, to carry canoes and equipment.' The men memorised who coordinated lunch preparations, who cut up the meat, who sliced the bread, who carved the watermelon and served it. After lunch, the boys spent the rest of the afternoon placing rocks as steps to the swimming hole and building up a rockpile to dive from. Red, the aggressive boy on the bus, directed their activities, and the men noted how the rest of the boys seemed to follow his lead.

The next morning, the boys took the path that climbed up through the trees above the camp to the stone corral at the base of the Robbers Cave, which, with its spring and grass, was a perfect place for outlaws to water, feed, and hide their horses. By midday they reached the top of the steep climb to Robbers Cave. After the exertion of the climb and the build-up, Smut was disappointed at how small the cave was. He was expecting something much bigger, but the cave was long and narrow, like a collapsed house of cards. Stone slabs were piled precariously on top of one another so that the cave's entrance was at the bottom of a steep slit in the rocks.

Bob Hood told them in his slow drawl how this opening was just the beginning, that down in the darkness the rocks had formed crevices and canyons and secret passageways that led to more hidden caves.

'Can we go in and have a look?' Red wanted to know.

Hood shrugged as if to say, *It's up to you.*

'C'mon!' Red called to the others.

One boy made his way closer to the entrance and wrinkled his nose at the stale air that wafted up from the cave. 'I'm not going in there,' he said, pushing his hands into his pockets as if to indicate that was that.

'Me neither,' another boy said.

Then a chorus of relief.

'Nope.'

'Not me.'

'Some other time.'

They were city boys and weren't quite brave enough to go clambering around in the dark.

But Red seemed to take their reluctance to go in as a challenge. 'I'm no yellowbelly,' he said. Even though he was the biggest boy, and he wouldn't have been able to squeeze into some of the smaller spaces that someone like Hollis would be able to negotiate, he took Hood's torch and disappeared inside the rocks.

Meanwhile, the rest of the boys climbed the rock above the cave and stood on the edge of a precipice that jutted out over a valley of pine trees, cooeeing and yelling, listening to their smaller- and younger-sounding voices echoing back from the other side.

They were getting impatient by the time Red emerged ten minutes later. Hood noticed how the boy had created a tension within the group but also how, after this trip inside the cave, he set the pace. If Red did something, such as climb a particularly high rock, Hollis was right behind him, and most of the other boys would do the same. They might have been too fearful to climb down into a dark cave, but in the sunlight they could pretend to be brave.

Unbeknown to the boys, as they were climbing the rocks above the cave on their second day, far down below another bus had arrived. A second group of boys who thought they had the place to themselves were disembarking, with Jack White, and carrying their things to a stone cabin on the other side of a small hill almost a mile away.

The men kept the two groups ignorant of each other during the first three days, making sure their paths didn't cross. It wasn't difficult: the two cabins were over a mile apart, at opposite ends of the camp. A pretty mountain stream flowed through the park, ending at peaceful Lake Carlton, where they could use boats and canoes. But the stream, named the Fourche Maline by early French explorers, translates into English as 'treacherous fork'.

Taking the lead on the research team was a challenge for OJ. At night, the men convened at 9.00 pm in the makeshift office that doubled as OJ and Sherif's sleeping quarters. A fan usually creaked on the table, stirring the hot air and ruffling the notecards OJ had laid out on the desk, detailing their hypotheses. 'I wanted to go through it methodically, let Jack and Bob offer their observations and evidence for each hypothesis from what they had observed during that day, but Sherif was eager. He was so eager that he practically jumped on them when they came through the door and chided them if they were even a few minutes late.' OJ shook his head. 'I had to get him under control. Jack and Bob were determined to do a good job — they were under pressure, as we all were, and Sherif was overstepping the line.' OJ was determined to remind Sherif of the 'ground rules'.

Did those ground rules include the fact that OJ was in charge?

OJ rolled his eyes and nodded. 'They sure did.'

How did Sherif react, I was curious to know.

'He acted upset, raised his voice, that kind of thing. But he knew I was not one to be bossed around. And he knew I didn't like to see him doing it to others.'

I suggested that with his doctorate completed and his position at Yale on the horizon, OJ might have found it easier than he had the year before to stand up to Sherif. But he shook his head. 'You'd think so, wouldn't you? But no, it was not easier. It was an uncomfortable feeling — I was always waiting for when he would blow up next. It hung over me and I couldn't relax. I had a deal of trouble sleeping.'

Hood took his group away on a two-night camping trip so that White's newly arrived boys could have the run of the park and spend their first few days swimming, playing ballgames, and taking their turn exploring the cave and its surrounds.

Hood and the boys hiked to a site four miles away. When they set out after breakfast, the air was velvety. The colours of the bush were still bright and the shade deep. By the time they reached their camping spot, the sky was a hard blue, and they were itching with dried sweat. The boys had to build latrines, put up tents, chop wood, and cook their own lunch with little help from the adults, all with the aim of cementing friendships. It sounded more like army life than a fun summer camp to me, but the boys seemed taken with the novelty of it all. 'I was a city boy from a poor area and I hadn't been in that kind of environment before,' Smut Smith remembers. 'I had stayed on my Aunt Leela May and Uncle Harley's farm in Texas, but this wasn't like a farm. The Robbers Cave wasn't like anywhere I'd ever been before. It was exciting. We spent a lot of time swimming and hiking and exploring. It was a real adventure.'

In the late afternoon, the men took them out canoeing. Bill Snipes and Smut were in Bob Hood's canoe, and the boys took turns paddling across the lake, with Bill keeping up an excited commentary on everybody's stroke. As they moved close to the reeds, they heard a commotion. Two snakes had hold of a large frog. Each snake had one of the frog's legs in its mouth, and they were thrashing about in the water. The boys stopped paddling, drifting as they watched in fascinated horror.

'Look at that!'

'Whoa!"

A loud crack and a boom made them jump. Birds exploded from the trees, shrieking. They turned to see Bob Hood aiming a gun and firing, *boom*, a second time. When they turned back, a spray of water was shooting up where the snakes had been, and the frog was gone. The sound seemed deafening, and it rang out and echoed back at them across the water. The boys looked at the quiet man sliding his gun back into a holster with a combination of fear and awe. Bill and Smut had never seen a real gun before, let alone a man shooting one.

That night around the campfire, Red and another boy argued about whether the snakes were water moccasins or copperheads. Hollis, a small boy with the authoritative air of someone who spent a lot of time around adults, announced they were rattlesnakes and suggested the boys should name their group after the snakes. Then he got them to vote on a name.

Hollis' pattern of resolving quarrels by vote and shortcutting disagreements made him popular with the other boys. Red, who tended to use his size and his fists to get his way, wasn't well liked by the others, and the more he sensed it the more aggressive he became. The boys were torn between their liking for Hollis, who was jockeying with Red for the role of leader, and their desire to appease the bigger boy. The men noted how Red had 'a pronounced tendency to rough up the smaller boys' for his own amusement and how they resented the way he took over and allocated jobs. When Hood suggested the boys would need to dig a latrine at their hideout, Red 'handed the shovel' to one of the smaller boys, who later complained to staff, 'We're tired of just doing the things he leaves over.' But the boy was too intimidated by Red to say it to his face.

Hood noted with satisfaction that by the end of the two-night camping trip a 'toughness' norm had emerged among the group, and gave plenty of examples to illustrate it. The boys' games grew increasingly reckless, for one. At a reservoir, Red instigated a game of climbing the slippery dam wall, but made no effort to help the others up it. Three boys slipped, and Smut remembers his skinned knees and the combination of water and blood flowing down his lower legs. Despite these injuries, the men noted gleefully, the group frowned upon crying. It was Red who so often set the pace in games of daring, and under his influence none of the boys wanted to be seen as weak.

But for all their supposed toughness, the boys were happy to return to the familiarity of base camp in the late afternoon of the fifth day. That night, they had a campfire at the stone corral, where they told jokes and sang songs, accompanied by Bill on his ukulele. Hollis hung a hand-painted sign on their cabin door: *Home Sweet Home*.

Yet this was about to change.

While Hood's Rattlers had been thrown in the deep end, left to deal with aggressive Red and the physical exertion of building latrines, pitching tents, and carrying water, back at the campsite Jack White's group had been having an easier time of it. For the first few days, Davey Munroe was the leader of White's boys. OJ had chosen him as a subject because he was a good athlete, but also, as the eldest of four, Davey was a good organiser. He had expressive black eyebrows that he wiggled as he spoke in a Donald Duck voice to cajole the others and get them on board, and the others looked to him to take charge in the first few days of camp.

By the second day, they had found a canoe near their hut — deposited there the night before by OJ and Sherif — and carried it downstream to a swimming spot. They were led by their junior counsellor, Bert Fay, who had recently fallen in love with geology (and would later make it his lifelong career), and along the way he pointed out fossils and interesting rock formations to the boys. In the water, the boys spotted a water moccasin and dubbed the place Moccasin Creek.

After their swim, the boys decided to build a rope bridge across the creek, and Davey coordinated the effort. A rather serious-looking boy called Will impressed them all with his knowledge of knots and the methodical way he measured out the spreader ropes. Dwayne, a skinny kid whose bony ankles stuck out from his too-short jeans, swam across the creek to tie the anchor ropes on either side. The job took most of the afternoon, and when it was finished, each of them crossed the creek, with boys on either side yelling advice as the bridge swayed over the rushing water. A chubby boy had trouble with the crossing, and he wobbled and several times looked like falling in. He got particularly loud applause when he made it to the other side.

'You're too fat,' Dwayne told him when he stumbled off the bridge among the backslapping of the other boys. 'You'll break that bridge.'

'Shut up,' a few of the boys cried. Davey Munroe scowled at Dwayne and turned away.

'I was an awkward kid, I always had my foot in my mouth,' Dwayne told me on the phone. He had the confident voice of someone who had left that persona behind long ago. 'My mother was always warning me. I used to say the first thing that popped into my head.' His laugh was a sharp bark. 'It was like I had no filter.' His mother was always giving him advice about how to get on better with folks since they'd moved to Oklahoma City from their farm down south, he said, after his father got a job building a new turnpike.

Dwayne tried to make up for it the rest of the afternoon by making himself useful. He volunteered to hold the anchor ropes to keep the bridge steady when the boys wanted another turn walking across, and he built the campfire and got it going when the others had trouble.

Late in the afternoon, Virgil, the large boy, yelled out, pointing at a copperhead sliding towards them, just eight feet from where they were sitting. All the boys except for Dwayne huddled together as Jack White and Bert Fay hit it with a rock and killed it. 'It didn't bother me,' Dwayne said. 'I used to go hunting with my uncle, so I'd seen worse.' He got them a stick to lift it up with so they could carry it off the path.

That night, while the boys were roasting potatoes and cooking hamburgers, junior counsellor Fay told them about the animals that used to be so plentiful in these parts, including bears and buffalo and mountain lions. After they'd gone to bed, the wind blew up and the boys listened to the rustling in the bushes outside. Perhaps it was the talk of mountain lions or the memory of the copperhead, writhing and bloodied, as the men pounded it with a rock, but one boy said that he missed his folks. Another boy started to sniffle. Will's voice trembled: 'They didn't say anything about snakes.' Davey told them all to settle, and the tent went quiet. In his sleeping bag, Dwayne silently prayed that the boys would like him better tomorrow.

# 8
# Nation States

In the men's office that night, when Jack White reported that two boys in his group were feeling homesick and a third was teetering, OJ Harvey sensed the change in Sherif's mood like a change in the weather. Sherif leaned back in his chair, his head tilted towards the ceiling. He made a tutting noise that OJ recognised as a sign of disapproval.

OJ shook his head when he told me. 'Oh lordy, if you think how much was involved — selecting and observing the subjects at their schools, going to their homes, visiting camps, arranging the buses. I'd spent a whole year on the logistics,' OJ said. In total, he'd spent more than three hundred hours choosing the final twenty-two boys. And it seemed as though he had failed to exclude the types of boys who would get homesick. By now, the chair legs had hit the floor with a thud and Sherif was up out of his chair and pacing, his face like 'thunder'.

'Jack, go call the parents,' OJ said quickly. 'Tell them we'll be bringing those two back home tomorrow.' Then OJ plunged on with the next item on the agenda: plans for how they would let the two groups of boys know the next day that they weren't alone in the park. What did Dr Sherif want them to say to the boys, OJ asked, and Sherif was soon distracted. But OJ could sense the tension in the man's shoulders. Later that evening, Sherif would want to go over and over this fact of homesickness and where OJ had gone wrong, talking long into the night. OJ realised then how tired he was. They all were, and it was just the beginning of the study. 'I had to nip it in the bud.'

At first I thought OJ was talking about his swift action to send the boys home, but he was talking about Sherif. That night, when the meeting was over and the only sounds were the whirr of crickets and the burp of frogs along the creek, OJ got in the truck, let out the brake, and rolled it down the road before starting the engine. He drove for eight miles, along dark country roads, until he saw the small can by the roadside, the secret signal that marked the turn-off, where he pulled in, switched off his lights, and bounced the truck slowly along the track until he came to a lean-to lit up by the light of a glowing fire. Then he made the reverse journey, this time with a bottle of moonshine.

OJ laughed at the memory. 'I had to go get that bootleg whisky every night. He was under a lot of pressure. We both were. Back at camp I'd fix him up with moonshine, and away we'd go.' The moonshine, much stronger than store-bought whisky, took the edge off Sherif's tension until the next morning. It was the only way either Sherif or OJ could get any sleep.

The next day, the Rattlers heard the sounds of the other group playing on the baseball diamond and, according to the men's notes, they reacted with dismay and immediately told Bob

Hood they wanted to challenge the interlopers to a game. But I wondered at this aggressiveness of the Rattlers, and whether the searing heat and the tension of keeping up with Red the bully had made them irritable.

In contrast, when Jack White told his group that there was another lot of boys in the park, they were enthusiastic, pestering him about meeting up to play ball. Back at camp, in preparation for the ballgame they had chosen a team name — the Eagles — and stencilled it on hats, t-shirts, and a flag the men had provided. The Rattlers group did the same. 'They were like little nation states. Both sides labelled everything as "ours" — "our creek" and "our mess hall". They pretty much claimed the whole place as their own,' OJ said with a chuckle.

As early as the second day of the camp, the observers' description of each group had already started to take a particular shape. The Rattlers 'conformed to a tough norm', and the observers' notes reflect a view of them as brave. When they hurt themselves — one burned his hand with fireworks, another dropped a heavy rock on his toe — they didn't cry or complain but 'cursed instead', the book the men later wrote about the experiment reported. The Rattlers were so determined to be stoic that 'staff had to remember to check' one boy's injured wrist and knee 'because he never mentioned them'.

Smut remembers hurting himself in a dam-climbing game. 'We would climb to the top of the dam wall and then slide down on our tail ends. I cut my leg pretty good. The men doctored it and bandaged it up.' But it was painful, and when Smut cried, Hood wrote 'he did not conform to this tough norm, and was completely ignored' by the others in his group. Along with the 'tough' norm went a definite approval of cursing. Hood described how one night, two of the smaller Rattlers pulled out their baby teeth together, and Hood cited this as further evidence of the 'toughness' that had become the group's rule.

The men described White's Eagles as less masculine but more caring towards one another. After the two boys were sent home, the group developed a rule against homesickness, and they were scrupulous about taking turns to say grace. And while 'swimming in the nude' became standard practice, those boys who couldn't swim stayed back at camp to fix lunch for the others. Unlike the Rattlers, who had named themselves in honour of the shooting of the snakes, most of the Eagles were frightened of the captured snake, and had given themselves a group name only after they heard there was a second group in the campsite, who they had wanted to play with instead of fight. The Rattlers dubbed their largest boy 'Red' for his 'size and toughness', but nicknames in the Eagles were 'Nudie' (for the boy who had instigated swimming with no bathers) and 'Marilyn' (for the boys who, during skits and songs, entertained the group by doing a 'burlesque dance', imitating Marilyn Monroe).

White made no direct comment in his notes about the Eagles' machismo or lack thereof, but OJ recalled them distinctly as 'crybabies', with none of the bravado of the Rattlers. 'A Rattler couldn't cry,' OJ told me. 'And the Eagles, they were cissies.' He chuckled. 'How on earth do these different norms emerge?'

It seemed to me that the personality of the group's leaders had a lot to do with it. The Rattler boys acted tough, and they cursed to appease the domineering Red. The tender-hearted Eagles valued fairness, and their leader, Davey, had been swift and clear in his disapproval of Dwayne's tactless comments about other boys. Perhaps the Rattlers were acting tough to hide their vulnerability, and the Eagles were praying as a way of holding on to the values of home in a world where things were new and strange.

On Thursday 24 June, the fifth day of camp and the day each group heard there was a second group in the park, the men organised the first of the 'frustration exercises', designed to foster solidarity in one group and stir animosity in the other. It was Eagle boy Davey's eleventh birthday, and that evening OJ and Sherif organised a surprise birthday party to 'cement group feeling' among the crew that had already lost two members to homesickness. The cook had baked a cake, and the boys crowded round and sang happy birthday to Davey, who blew out the candles in one long breath. Davey asked Jack White if he could invite the other boys. But in order to 'stir animosity' in the other group, White replied that the others were busy and there wasn't enough cake to go around.

After the singing and the cake, the boys played charades. Across the creek, the Rattlers sat around their campfire listening to the singing and the hip-hip-hoorays, excluded from the fun. It reminded me of a story I'd heard when visiting Alcatraz: how the worst night of the year was New Year's Eve, as when there was a southerly wind the prisoners in their cells could hear the sound of music and laughter and the tinkle of glasses blown across the water from the San Francisco Yacht Club. I could imagine how the Rattlers must have felt: cast in the role of outsiders, the boys likely sat there feeling a mixture of resentment and loneliness.

The next morning, Sherif's voice was both solemn and excited as he leaned into the microphone and told the tape recorder that the team had hidden in the mess hall to record interactions, 'Today, June 27, 1954, we have started Stage 2 of this experiment. There is going to be a series of contests or a tournament. Knives and medals will be used as rewards ...'

The announcement of the tournament was a big day for the men. As with the competition in the Middle Grove study, it marked the end of the first stage of the experiment, which they called 'in-group formation'.

OJ made the announcement to the Rattlers first, and then to the Eagles. When the Eagles had finished clearing away their dishes, OJ proclaimed that instead of just a ballgame, the counsellors for the two groups had gotten together and organised a 'nice surprise' tournament with 'some fine prizes'. I imagined Sherif leaning in the doorway between the kitchen and the hall, sipping a cup of coffee and feigning a lack of interest, but his eyes following every move, every facial expression and verbal reaction of the boys. He would have nodded to himself approvingly as OJ made Sherif's careful script sound like a casual announcement.

Yet as OJ outlined the activities involved and the point-scoring system, the Eagles were restless; they seemed overawed, rather than excited. Some, like Dwayne, were bothered by the extravagance of the prizes — a ribbon and a jackknife for each boy, as well as a trophy for the winning team, and nothing at all for the losers? On the tape you can hear the worry in their voices at the idea of such a serious competition. Perhaps Davey was concerned that the other boys were mad they hadn't been invited to his birthday party. 'Maybe we could just make friends with these guys,' one boy said to OJ. Some of the others agreed that it didn't seem fair not to have prizes for the losers, and that the group who lost the tournament were likely to be resentful. 'Someone is going to get mad and hold a grudge,' Will said nervously.

To OJ Harvey and Jack White, the boys' reaction was further evidence that the Eagles were cissies rather than toughies. 'They were timid,' OJ said, 'they weren't rearing to go like the Rattlers.' But I imagined some gesture of impatience from Sherif prompted Jack to step in. 'Wait up,' he said as the boys got ready to leave. 'These here are real expensive knives, the best steel that can be bought in McAlester.'

'Listen to Jack,' OJ admonished them.

But the boys were eager to leave. Suddenly the friendly ballgame they had suggested had been transformed into a full tournament, including three baseball games, three tug-of-war contests, a game of touch football, three tent pitchings, daily cabin inspection, a skits-and-songs contest, and a treasure hunt — a total of sixteen events, scheduled over four days, in 100-degree heat. No wonder they were eager to get back to their swimming hole to play and cool off. A full-on tournament where the winner got all the prizes didn't seem fair. And despite OJ's attempts to convince them that it was an unexpected bonus, it must have sounded more like hard work than fun.

Not that I could find any mention of the Eagles' reluctance in the book the men later wrote, *The Robbers Cave Experiment: intergroup conflict and cooperation*. The book describes both groups at the end of this stage as rearing to compete. They were 'unanimous' and 'insistent', so much so that 'delaying the contest became increasingly difficult', the men wrote, as if they had to hold the boys back.

The book describes the experiment's three stages as a smooth narrative. At the end of this first stage, the book stated, each group had a name and a 'definite group structure'. They had flags, a symbol of their identity; they had group norms, or shared ways of doing things, like the nicknames they gave some boys; they had taken 'ownership' of territory. But how much of a hand did the men have in this remarkably neat unfolding of events? In the book, the researchers were invisible, their role kept carefully backstage. The story about how one group of boys named themselves the Rattlers in honour of their gun-slinging leader is not included, nor does it mention the party and games that favoured one group and fed the antipathy of the excluded boys.

The audiotapes, the handwritten notes, the memories of some of the boys I spoke with, as well as OJ, create a more complex picture. What else was being left out, I wondered, and who, if anyone, was playing the role of scientific conscience of this experiment and making sure the men were maintaining some kind of objectivity?

Towards the end of the first stage, Sherif wrote to his wife:

> ... My Carolyn — your letter and your phone call yesterday brought a warm touch to our hectic but exciting life here which is always on the move in various colorful and ever changing ways. One minute we'll be attending to the drainage of boys' showers here, the next moment it may be a highly abstract problem in group relations ... We are in a fascinating transition stage now. One group adopted the

> name of Rattlers last night at the Robber's cave during their camp fire. The other group named their swimming place, camp fire, and cabin … The great likelihood is that stage II will start to-morrow. The boys of one group are busy now 100 yards away from here putting their … name … on their shirts and caps …

If the two groups of boys seemed to have melded in the first week of the camp, Sherif and his fellow researchers seemed to have too. Gone was any mention in Sherif's letters of the desperate loneliness he'd felt the year before or the melodramatic view of himself struggling alone under the yoke of a 'Herculean' task. In his letters, the first person 'I' had disappeared, and he refers constantly to 'we'.

But OJ knew that Sherif's jubilance could evaporate as quickly as it came. And also, at the first sign of a problem, how quick Sherif was to lay blame on one or all of his staff. After the announcement of the tournament and prizes failed to get the Eagles enthused, Sherif had begun to hector Jack White for not doing enough to get them riled up. OJ found that keeping Sherif busy helped. Despite his lack of practical skills, Sherif got great satisfaction from acting the role of handyman. As well as covertly observing, Sherif helped OJ with the demanding practicalities of the camp and arrangements that were often last minute. They had the broad outline of the three stages mapped out, but exactly how each group would spend each day, and what equipment and supplies they needed, were all decided the night before.

'Our planning was one day at a time, for the most part. It was very intense,' OJ told me. Sherif helped OJ unload equipment and supplies from the truck, prepared the baseball pitch for a game, unblocked drains, chopped wood, and delivered food for cookouts, where he and OJ lingered to observe preparations. They tried not to arouse the boys' suspicion. 'Early on, we started using a camera all over the place so the boys wouldn't be bothered by cameras. And we tried to keep the note-taking out of the boys' view. It was a task all right,' OJ said.

But obviously they hadn't managed to keep themselves completely out of sight. Sixty years later, Smut Smith told me, 'I wasn't surprised when you told me it was an experiment because I do remember the men had notepads that they wrote on. I wasn't suspicious at the time, but I do remember that vividly.'

Smut's observation of the men taking notes suggests that the research team had learned little from the failed 1953 study about the curiosity of the boys or that the observers themselves were subject to the boys' scrutiny. Despite the men's description of apparently scrupulous and surreptitious methods of recording, it was obvious that there was a gap between the ideal of an experiment as it was described in their publications and the reality of its execution.

On the day the tournament began, the Eagles voted Will captain of their team even though Davey was their unofficial leader. Will was their best ball-player. He applied the same methodical and patient approach to games as he had to the knot-tying on their rope bridge. The men were pleased with the choice because they felt it would likely head off Will's tendency to homesickness. Davey Munroe took the vote with good grace, pounding Will on the back in congratulations.

In their cabin, Will gave the group a pep talk about staying focused during the game. The nine boys set off in their new t-shirts, with Davey carrying their flag, which featured a picture of an eagle with its wings outspread. They marched through the park and crossed the small creek and climbed up onto the baseball pitch, looking forward to meeting the other group of boys. When they arrived at the pitch and saw eleven Rattlers lined up and staring at them in sullen silence, the Eagles went quiet.

In photos, the Rattlers were strung out in a long line, watching the new group warily as they approached. I guessed that it was Red who was waving a flag with a large black snake on it and scowling. Hollis stood with his hands in his back pockets, sizing them up as the Eagles boys approached.

Perhaps Sherif sat on the sidelines with his back against a tree, his cap tipped back as if he had casually settled in to watch the game.

The Eagles were unnerved by the unfriendly reception. Davey let the flag droop and trail in the dirt. Once the game started, the Eagles were alarmed by Red's cursing, and that the umpire did nothing to rebuke him. Red jeered, calling them 'dumbasses', and Virgil, on bat, 'Tubby' and 'Little Black Sambo', and a few of his teammates joined in. Davey defended Virgil and retaliated by calling one of the Rattlers a 'retard', but once the game entered its middle stages the insults died down.

At the end of the game, which the Rattlers won 16–14, they gave three cheers for the Eagles, a fact that Sherif scribbled in his notes, underlining the phrase 'Exhibition of good sportsmanship on the whole (Am.norm)' as a reminder of something to watch out for.

Back at the cabin, Eagles captain Will berated the boys for ignoring his earlier instructions and getting distracted by the tactics of the other team. They resolved that in future games they would ignore any cursing from the Rattlers and not engage in it themselves.

That night, after supper, the tug of war started. Each team lined up on one side of a heavy white rope, which Dwayne remembered was rough and scratchy in their hands. Straining at the rope, the Rattlers pulled one Eagle after another over to their side. Afterwards, they marched off together, singing, 'The first Eagle hit the deck, parley vous, the second Eagle hit the deck, parley vous!' all the way to their cabin on the far side of the park.

I imagined their voices carrying through the twilight, the sky streaked with violent pink as the sun went down, and a man stepping behind a nearby tree to listen and watch.

The Eagles stood around the white rope now, tumbled in the dirt. Davey kicked at the rope. 'They're much bigger than us.'

'They're eighth-graders — they must be, for sure!' Will's voice trembled and he looked about to cry.

Will turned to Davey. 'And you! Why'd you drop the rope? You dropped the rope!'

'We were beat,' Davey said, shrugging his shoulders in a gesture of hopelessness.

The book reported the conversation above, naming each boy's role in the episode that followed. On their way back to the cabin, the dejected Eagles spotted the Rattlers flag, which had been left on the backstop. They pulled it down and tried unsuccessfully to tear it. Then 'someone' suggested they burn it, the book notes. Who was the unnamed 'someone', I wondered? And who gave them the matches? The boys lit the flag and hung the burnt remnant on the backstop before running off.

'The burning of the flag was exactly like a declaration of war,' OJ said to me.

Of course, there was retaliation when the Rattlers found the remains of their flag the next morning. They swarmed towards the Eagles, and as they approached the pitch, shouting that the Eagles were 'bums' and 'little shits', some of them rushed forward and seized the Eagle flag out of Davey's hands and ran off with it. Red grabbed Davey in a headlock and wrestled him to the ground, and they rolled in the dirt until Jack White and Bob Hood rushed onto the pitch to break it up.

But the fight electrified the Eagles. They huddled together before the game and, at Dwayne's urging, said a fervent prayer before playing baseball with fierce energy and focus. The strategy they'd agreed on the night before was to fight fair, to be good sports, and not to cuss or brag in front of the other team. Despite the provocation of the Rattlers before the game and their continued cat-calling, the Eagles didn't take the bait.

The Eagles were ecstatic when they won the game.

'See? Didn't I say so?' Dwayne ran round excitedly asking each boy in the team, attributing their win to their prayers. But Virgil said the Rattlers lost because they cursed so much. Still, the Eagles gave three cheers for the Rattlers, who had felt so confident of winning at the start and were now downcast: Sherif wrote in his notes they were feeling 'very low'.

In their book, the men noted, 'This flag-burning started a chain of events that made it unnecessary for the experimenters to introduce special situations of mutual frustration for the two groups.' He made it sound as if the men simply stood back and let the drama unfold. But it was far from that straightforward. Sherif was clearly bothered at the end of the game when the Eagles gave three cheers for the losing team. In his notes, he commented anxiously that the earlier friction had evaporated, and wrote that what he observed was 'the potency of the thoroughly ingrained sportsmanship norm'. He was troubled too because it raised the spectre of the failed 1953 study and the rules of fair play those boys had insisted on during games.

Back at their cabin after the loss, the Rattlers were turning on one other, 'dejected because of the loss and fatigued from heat and exertion … tempers were short and the bickering went from bad to worse,' Bob Hood wrote in his notes. Anxious that the boys were directing their aggression and blame towards their own members instead of to the other group, Sherif called a hurried meeting with OJ. It was time for another 'frustration episode' to bring the Rattlers together as a team and to kickstart conflict. But this time, the saintly Eagles would be the victims.

Close to midnight, the Rattlers huddled outside their cabin. That day, the Eagles had beaten the Rattlers in a tug of war by sitting down while they held the rope, which the Rattlers felt was 'unfair tactics'. According to Sherif, 'the mood was definitely favourable to a raid'. The Rattlers had painted their faces with soot from the fire 'commando-style', and the whites of their eyes glowed in the dark. Suddenly Hollis shot forward and they followed him, running, doubled over, across the patch of dirt between their cabin and the mess hall.

They moved as a pack across the creek and then paused again

on the other side. Hollis was whispering and gesturing. Bill shot out ahead and hoisted himself up to the window of the Eagles' cabin and threw himself against it, falling through the screen with a crash. A howl went up, and a blinding light went off as Bob Hood, supposedly being surreptitious, snapped the first photo. Boys yelled and shouted and whooped. Some boys ran around the building in circles; others ran inside. Red was yelling, 'Eagles are chickens! Eagles are chickens!' Inside someone was crying. Two boys had jumped out of bed and were running towards the doorway, waving their arms. Red shook the frame of one of the double bunks and, as two boys huddled under their blankets, he whooped and howled, picking things up and throwing them — books, towels, and other possessions — before he followed the others out again, where they pounded across the park in the dark.

It was all over in a few minutes, but in the Eagles' cabin it was bedlam. No one could find the lanterns, so they couldn't see. Someone was sobbing. Will was yelling 'You yellowbellies!' but no one was sure who he was yelling at, the Rattlers or the boys, like Virgil, who had stayed in bed, pretending to be asleep, until the noise and shouting was over. Jack White appeared in the doorway with a flashlight. The boys went quiet as White played the light across the trampled clothing, the books and cards spilled on the floor. Then they all started talking at once. Someone said they should go get the other boys and retaliate, but White said no. So what was White going to do about it? Will demanded. The Rattlers should be disqualified from the tournament for being such sore losers, Davey chipped in. White said he'd talk to OJ Harvey. In the meantime, Bert had arrived with some spare lanterns and the boys decided to clean up straightaway, rather than wake to the mess in the morning. As they shifted their beds and picked up the clothes, they went over and over what had happened. Virgil and Davey took a lantern outside to find some rocks so they could fill their socks with them and use them as weapons in case the Rattlers came back.

'We didn't like that other group, that's for sure,' Bill Snipes said. He had just retired after more than forty-five years in the Oklahoma police force, first as an officer, then as a detective.

'We tried to outdo them, and then we started doing things to the Eagles group, raiding their cabin, messing up their stuff. It started with the tug of war,' he continued loudly. He was taking me for a spin around Oklahoma City in his old Jaguar. Low to the ground, it roared and rattled, which at the best of times would have made it too noisy to talk, but was even trickier given Bill was hard of hearing.

'I remember we called the Eagles group a lot of bad names. It was almost like the counsellors were building this animosity. Then we raided each other's cabins. I climbed through their window and almost fell on Davey Munroe. I woke him up and he was not happy. He started swinging at me. We tore their place up. They did the same to us. But I don't remember who started it.'

Bill might not remember who started it, but it was the staff who kept the animosity going. The boys clearly looked to the men to police the misbehaviour of the rival group. The next morning, when White visited the Eagles in their cabin, there was no longer any talk of revenge, and they told White that the Rattlers should be disqualified from the tournament. But White suggested they could even the tournament score by dumping mud in the Rattlers cabin while the other group were at breakfast. That way, the Rattlers would lose cabin inspection and the Eagles would take the lead.

Returning from breakfast to find buckets of dirt upended in their cabin and their clothes and bedding on the floor was all the provocation the Rattlers needed. By the afternoon's touch football game, any sign of guilt on the part of the Rattlers was gone. They marched to the ground with Will's jeans, now emblazoned with orange paint reading 'The Last of the Eagles', swinging on a new flagpole that they carried side by side with their Rattlers flag.

That week of the tournament, each day was like a tug of war, with the barometer in the mess hall inching up, putting the Eagles ahead one day, the Rattlers the next. Manipulating the scores in games was too risky, but there were other ways to control the scoreboard. 'We cheated a bit. We wanted them to be neck-and-neck right to the last, so we created things like cabin inspection and songs and skits so we could keep the points close,' OJ said.

On the night before the end of the tournament, White argued forcefully for making the Eagles the winners. 'They were less sturdy, their morale was already low, and we figured if they lost, the study would be over. They'd want to go home,' OJ remembered. 'We knew the Eagles would fall apart if they lost. We knew the Rattlers were a stronger group, they could handle losing.'

When Will had seen the Rattlers parading his stolen jeans around the football ground as a trophy, his homesickness had returned. Virgil, who the Rattlers had continually mocked during the contests as 'Fatty', had also written home, saying how much he was missing his folks. Sherif regarded the name-calling, the skirmishes, the raids, and the retaliation as proof that competition leads to hostility. But the breakdown of the relationship between the boys carried risks, and to lose any more boys from the already depleted Eagles team now, after getting this far, would be a catastrophe. Sherif wrote in his notes, 'We couldn't take a chance on the Eagles getting dis-integrated.' OJ and Sherif decided to take White's advice.

The scores were so close at the end of the five days that the final event, the treasure hunt, would be the decider. The team who found their treasure fastest would win the tournament, and the knives, medals, and trophy.

Both groups were quiet and fidgety as Bob Hood made the final calculations on his clipboard. 'Now, for the results of the last contest, the time for the Rattlers was ten minutes and fifteen seconds. The Eagles' time was eight minutes and thirty-eight seconds, which gives the Eagles the —' I couldn't hear the rest of Hood's announcement on the audiotape for the screaming and cries that came from the exultant Eagles, who, the observer notes told me, were jumping and hugging one another. Will 'cried with joy' and Dwayne danced around holding the trophy, with each Eagle taking a turn to kiss it while the Rattlers were 'glum, dejected, and remained silently seated on the ground'. The tape didn't record what happened next and nor do the observers' notes, perhaps because it was an unwelcome spectacle. The men wrote in their book how during the tournament the 'good sportsmanship' norm 'gave way' to 'hurling invectives' and 'derogation of the out-group'. The Eagles gave three cheers for the losing Rattlers, a fact Dwayne doesn't remember but doesn't find surprising. 'We would have been so relieved it was over,' he said. 'The end of the tournament meant the end of the fighting.'

The last thing Sherif wanted was any spontaneous gestures of reconciliation between the groups. The men had an impromptu meeting immediately afterwards, and in it, tensions ran high. This was the point in the study the year before when things began to fall apart. OJ told me how in the office Sherif jabbed his finger at White, accusing him of jeopardising the research, of not doing enough to fan animosity. White said nothing. One of his favourite sayings was 'He who shouts first loses the argument', his daughter Cindy told me, but he and OJ had exchanged a look.

OJ looked out the window in his Boulder home, at the mountains. Jack White had died in 1988, and OJ still missed him. 'Jack was the dearest friend I ever had. We went hunting and fishing and riding and telling lies about the fish we caught and drinking …'

I gave him a moment, looked away.

He cleared his throat. 'Where was I?'

'I was angry, but I took care not to show it,' he continued. 'I said, "Now, Dr Sherif, you remember our agreement. You don't do that."' And he looked from me to Jack and back again. He knew that if I wasn't allowed to be in charge, Jack and I would both quit. I didn't have to spell it out, but he understood it. Later, he apologised. But it came close. It came close to being ugly.'

OJ had averted an emotional showdown this time. But what struck me about the scene that he described was what it revealed about power. Contrary to Sherif's own theory, the leader of the group was not the person with the most status but the one with the most power to derail events or deliver the study they had in mind.

While the boys would have thought the tournament and the conflict was over, the men had other plans. The third and final stage of the experiment — the gluing together of the rival groups — could only work if the relationship between the Rattlers and the Eagles was completely broken.

That afternoon, while the Eagles were celebrating their victory at the lake, the Rattlers raided their cabin and stole their medals and knives, tearing the insect screens off the windows, overturning beds, ripping up comics and, down at the creek, setting the Eagles' boats loose. Sherif's notes read that 'a very destructive job was done … staff had to put the brakes on'.

When they returned and saw what had happened, the Eagles, headed by Will, raced across the park to the Rattlers' cabin. It was the scene of a 'most dramatic' fight, Sherif wrote. Will confronted the Rattlers. Red said they would return the knives if the Eagles got on their bellies and crawled. Will demanded that he send someone his size out to fight, but Red just laughed. Will was 'hysterical' and rushed forward, and all hell broke loose. 'They had thrown each other down and were slugging it out and I had to get in and break it up,' OJ said.

With some trouble, the men separated the boys, and OJ shepherded the Eagles back to their cabin. Will's voice began to shake as he recounted to OJ how the Rattlers didn't want to fight fair — there were eleven of them and only nine of the Eagles — and how the counsellors should have taken out those two big boys so the groups were matched. Perhaps it was the sympathy in OJ's voice, or the sight of the knives and the medals that he had returned to them, or that the adrenaline and anger of the fight was wearing off, but the boys' earlier bravado evaporated. Sherif, standing outside the cabin listening, described in his notebook how the Eagles '... cried and sobbed' and were 'confused and insecure'.

'The end result of the series of competitive contests and reciprocally frustrating encounters between the Eagles and Rattlers was that neither group wanted to have anything whatsoever to do with the other under any circumstances,' the book states. But OJ was worried that things had gone too far. The emotional distress of the boys, and in particular the Eagles, could create problems with 'public relations' — a term he and Sherif used for complaints from parents and action from the university. That night, he and Sherif decided that the boys needed time to recover before the final stage of the experiment began. They kept the two groups separate the next day and took them out to boat and swim.

Giving the boys a day of fun in their own groups was also an attempt to avoid the kind of complaints the Rockefeller Foundation had recently fielded about Sherif's group study, concerning his and his team's treatment of the children.

In December 1953, a number of small-town newspapers ran stories about Sherif's Middle Grove study, based on reports from a conservative journal called *Human Events*. One editorial, under the sarcastic headline 'No (Public) Comment', stated, 'Dr Sherif deliberately, surreptitiously and secretly planted ideas in the heads of boys and played on their emotions in such a way that the friendly rivalry between the groups was converted into a hostile one. He also worked upon individuals in each group, causing each to brood ... causing grudges and resentments.' The notion of a foreign-sounding scientist preying on the minds of naïve children echoed widespread fears about vulnerable people falling sway to the power of communist brainwashing and propaganda. It also cast the Rockefeller Foundation's willingness to fund it in a poor light.

Leland C. DeVinney, the Assistant Director of Social Sciences at the Foundation, wrote to his superiors that the articles were 'sheer invention' and 'an astonishing distortion': 'fantastically distorted and hostile'. But at least one reader reacted, and wrote to the Foundation directly to ask for more information about Sherif and the study. Even without the implication that it was providing research support to America's enemies, the Rockefeller Foundation was sensitive to accusations about its funding of research with children, having been at the centre of previous controversies for its financial support of what was deemed unethical experimentation — as early as 1911, Rockefeller-funded bacteriologist Hideyo Noguchi had caused a public scandal after he was accused of infecting children with syphilis.

Upon receiving the letter, DeVinney wrote to Sherif asking him to supply information so he could answer the man's questions, in particular what exactly was it that parents had been told? When DeVinney answered the letter of complaint he wrote that the parents, 'while they did not know the study design', were 'told this was an experimental camp'. But what he didn't add was that Sherif had, he wrote to DeVinney, told the parents it was a 'study of camping procedures', not a study of intergroup hostility.

The exchange had made Sherif anxious, and both he and OJ were careful at Robbers Cave to keep parents in the dark about the real nature of the study. Throughout, they read all the boys' outgoing mail to make sure no news was going home that might cause trouble.

I imagine that the day's break away was a welcome relief for the staff too. The 1949 study had ended at this point, and the 1953 study failed to get the boys fighting. 'We had never reached this stage before, so it was unknown territory as far as we were concerned,' OJ said. 'We had real reason to be concerned about whether we could pull it off.'

# 9
# Sweet Harmony

Carolyn wrote excitedly to Sherif on the first day of the final stage of the experiment, telling him how tempted she had been to call. 'According to schedule you must be through with fighting (experimentally) and heading towards sweet harmony!'

If only it were that easy.

Before the final stage could begin, Sherif wanted to gather hard data to support his anecdotal evidence that the experiment was working by bringing both groups together to test their attitudes to each other. Bob Hood told the Rattlers he'd made a bet with Jack White that the Rattlers were a better team than the Eagles in certain skills tests. But by now the Rattlers didn't want anything to do with the others, and refused to participate. In desperation, Hood promised his group $5.00 if they took part. In a bean-tossing competition where the boys had to estimate how well their group did in comparison to the others, each group was much more positive about its own members and regarded themselves as superior in skill to the other team.

In the published version of the study, the men portray the

final week of the camp as an epic struggle to overcome the aggression and fighting between the two groups of boys. They described how the boys, waiting for meals outside the mess hall, abused one another, and once inside at the table, 'After eating for a while, someone threw something, and the fight was on. The fight consisted of throwing rolls, napkins rolled in a ball, mashed potatoes, etc. accompanied by yelling the standardized unflattering words at each other.' Name-calling and abuse were a feature of every interaction, according to the book. And when things went wrong, the other group generally got the blame: 'In an early morning swim … the Eagles had discovered their flag in the water, burned the previous evening by the Rattlers. Upon making this discovery, they denounced the Rattlers as "dirty bums", and accused them … of throwing rocks in their creek (because one of them stubbed his toes a number of times during the swim).'

It makes for uncomfortable reading. Each time the two groups came into contact, whether it was to eat a meal or watch a movie, the men made notes on how they acted, where they sat, and the overall mood between them.

'We were trying to show just being together doesn't create peace. You need more than that,' OJ said.

For the first day, mealtimes were bedlam. Boys on both sides 'yelled invectives' and threw food, the book says, while the men hid outside and took notes. On the second day, after joint meals in the mess hall, Ida the cook (who was wife of the camp's permanent caretaker, Dave Bloxham) and her sister (who worked as kitchenhand) threatened to quit. They were 'conservative churchgoing people', OJ said. 'She was so bothered by the misbehaviour of the boys, who were throwing food at the other group and cursing each other. The food fights and bad names really bothered the ladies, and they chided Sherif and I for allowing the boys to behave that way,' OJ said. Ida's husband Dave took her aside and explained. 'To our surprise, he had pretty well mapped out what we were doing. He saw that we were trying to create discord and break it down and he had a chat with his wife and sister-in-law and talked them out of quitting.'

My guess was that the boys were angry and upset not only because they felt aggrieved by the cheating behaviour of their opponents but also because they felt betrayed by the camp's staff. They were confused.

In their published accounts, it is as if the men have let a genie out of the bottle in this final week, but the reality was quite different. Boys on both sides were angry and upset. It made me wonder, how had things become so heated? And why were these boys at Robbers Cave so much more easily drawn into conflict than the boys a year earlier?

Dwayne Hall told me, in his gravelly smoker's voice, that he thinks it was because all the usual rules were upended. 'We — our group — we were playing by the rules. But that other team, the Rattlers? After a game, in the middle of the night they attack us and steal our stuff. That's like —' He can't find the word for it. 'Well, what do we do? We tell our counsellors because that's not playing fair. We're like, "You got to disqualify these guys." But what do the counsellors do? Nothing. Maybe they're the ones who suggest that we get our own back, maybe it was one of us, I don't remember. But there was no way that the fighting between us was "natural". It was crazy — a crazy situation run by crazy people!' He sounded angry.

What had the others said about the camp, he wanted to know.

He was the only one of the Eagles group I'd managed to find, I told him.

'Doesn't matter which group,' Dwayne said briskly. 'We were all brought up to play fair and square. You can't tell me that the boys in that other group didn't feel bad, didn't know what they were doing was wrong. I mean, there were these free-for-alls, like a riot or something. I mean, we felt under attack. We were filling our socks with *rocks* at night. So we were planning to retaliate. I mean, we had to look after ourselves because the counsellors weren't going to.'

Did the Rattler boys feel betrayed by the men too? They were certainly resisting Bob Hood after the tournament. His inconsistency must have been confusing and upsetting: how he accompanied the Rattler boys on raids but then didn't take their side when the Eagles clearly broke the rules during tug of war.

Bill Snipes doesn't remember much about the fights but is confident that they occurred. 'There was a lot of pushing and shoving. I guess we were pretty mad. Some kids were pretty upset. Our counsellor was with us all the time — they knew about it. How else would eleven-year-old boys hurt each other?'

On the second morning of the last week, when the boys lined up for breakfast, Ida came out from the kitchen and stood at the door to the mess hall to give them a piece of her mind. With all the extra cleaning up she and Ruth had to do because of the boys' mess, the two had less time to cook. If the boys wanted to eat, they had to behave themselves, the small woman with the neat grey bun told them firmly.

Was it a coincidence that after breakfast that day the Rattlers took a stand against their bully? Back in their cabin, Red, with his penchant for picking on the smaller boys, pulled the comic out of the hands of a boy called Franklin, who was reading on his bunk. Franklin leapt up to grab it back. Red shoved him and Franklin fell backwards, hitting his head on the railing. Then Red was on top of him, punching him as Franklin kicked and cried underneath.

Bob Hood raced in and pulled the boys apart. I imagined the freckles on Red's face stood out in bright blotches as he struggled free of Hood's arms and ran out of the cabin. Hood hurried after him. The boys crowded around Franklin, who was rubbing his head and trying not to cry. When Bob Hood came back and asked if the boys knew where Red had gone, Hollis spoke up: 'We'll bring him back if you stay out of this.' The boys drew in closer to stand behind Hollis.

The story is an awkward wrinkle in the smooth narrative of the experiment presented in *The Robbers Cave Experiment: intergroup conflict and cooperation*. But it's stayed with OJ, and he tells the story with a kind of wonder at the boys' solidarity. Clearly they had given up expecting the men to take Red in hand. I wondered if they had been inspired by Ida, who seemed the only one among the adults laying down any rules about right and wrong. Maybe it was Ida's talk that gave them a burst of courage.

Hollis and Bill searched the camp and the woods and returned with Red an hour later. The boys adjourned to the mess hall with the rest of their group, with no adults allowed. The men only learned afterwards what happened.

'They held a kangaroo court and kicked him out,' OJ said. 'Told him he couldn't be part of the group until he apologised.' But Red was stubborn. He returned to his hide-out in the bush.

Noon came and one of the smaller boys brought him a plate of food, 'like a puppy', OJ told me, and '[he] said, "All you've got to do is apologise and you can come back in," but Red wasn't having any of it. He spat at the boy's feet and said nope, he was staying put, he wasn't about to apologise.' Red sulked alone all afternoon, but that night after supper, 'he came limping into port and the boys made room for him at the table and continued with their card game. No one said anything. But he never bothered any of them again,' OJ said admiringly, shaking his head. No wonder the men left this out of the published version of the experiment, I thought. How would they accommodate this in their story of the united and aggressive Rattlers?

After lunch, OJ told the boys to fill their water canteens because there seemed to be some problem with the water supply and he needed volunteers to help find the problem and fix it. So began the experiment's final stage, where the men introduced superordinate goals or problems that the boys could only solve together, with the idea that this would restore relations between them. This time there had been no talk of setting a forest fire, I imagined because OJ was practical enough to realise it was a dangerous and unpredictable way to try to unite the boys.

In their groups, the boys, trying to find the problem, slowly traced the water line from the mess hall, up the mountain, to the water tank at the top, where that morning OJ and Sherif had buried the valve under a fall of rocks. Sherif made few notes on the day, busy following the boys up and down the line, pretending to be flummoxed by the problem and, like the boys, looking for a solution. At key moments, when the boys' backs were turned, he took photos. The men had their hearts in their mouths, OJ told me, watching the boys move along the line, trying to find where and why the water had stopped flowing. 'The whole idea was that they would have to work together to fix it,' OJ said. 'But we didn't know if it was going to work. We were improvising.'

At the beginning 'they were quite touchy about interacting', but slowly, with the sun beating down and their water canteens slowly emptying, the boundaries between the groups began to blur. I pictured the Rattlers, free of the tension of Red's bullying, were more relaxed than they had been for days, and the Eagles, anxious to put an end to the fighting, were ready to meet the others halfway.

When the boys finally found the valve buried under the rockslide at the top of the mountain, the groups took turns lifting and carrying the rocks away. But, realising there was a better and faster way of getting the job done, they soon formed a chain, passing the rocks down the line and working as a single team. In the photos you can see the boys standing on the ladder and clambering over the top of the water tank. I thought about how hot it was at Robbers Cave in July — how the metal from the water tank would radiate heat and would burn to the touch, the ground around it dry and dusty — but the boys clambered eagerly over the tank and stood peering up at it at different rungs on its metal ladder. They were clearly motivated, and pleased to be able to help.

By sundown, however, after three hours in the sun lifting and hauling rocks, the boys were exhausted. 'We'd done too good a job on those rocks,' OJ said with a laugh. 'In the end we had to solve it. They needed our help, so we helped them lift off the final rocks.' With the water supply restored, and as much cold mountain water to drink as the boys wanted, supper that evening was a relatively 'calm' affair. 'There was some joking between the boys, although things were still a little strained. But it was the beginning of the breakdown of the antagonism.'

Despite the others' tiredness, Sherif was full of energy. 'He was elated,' OJ said. 'We both were. We could have pulled those rocks off single-handed we were so pleased.' He laughed. 'He was an egalitarian guy. He was against competition because it creates prejudice and inequality. Now we had created a situation where groups would work together and break it down.'

But it was just the beginning, the tentative first steps. 'We didn't get ahead of ourselves,' OJ said. 'We had to come up with other things to get eleven-year-old boys to cooperate. We were saying, "What on earth are we gonna do next?"' You wouldn't know it from the published description of the final week that it was the result of last-minute planning.' The next day, the men announced that their budget didn't cover the full cost of hiring the movie *Treasure Island*, so the groups pooled their money; the day after, they went camping together at a lake. 'They got one tent up and they didn't have time to put the second one up before a very severe storm came over, as if we'd ordered it,' OJ recounted and chuckled. I imagine that sheltering in one tent for the night, with the excuse of the storm, allowed the boys a chance to make friends without loss of face.

The last event for the week was a trip to Arkansas. The boys were told there would be two trucks, with a group in each truck.

All of them were excited at the prospect of crossing the border into another state. The Rattlers were so keen that they wanted to leave without breakfast. But that would have spoiled the men's plan. After the boys had eaten, OJ announced solemnly that one of the trucks was out of action: 'We only have one truck fit to take campers to Arkansas so we might have to give up on the idea of going.' Over the boys' cries of disappointment one boy yelled, 'We can all go in one truck, OJ!' and the others joined in. Sherif noted that a couple of boys were silent and seemed resistant to the idea, but they were swamped by the general enthusiasm. It would have been near impossible for OJ not to grin, or glance over at Sherif, who I pictured hovering by the door, wiping his hands on a rag as if he had been looking under the bonnet of a broken-down truck.

It was a three-hour drive, and they stopped halfway, at the small town of Heavener, for a drink. By now Sherif and OJ were elated. While the boys drank sodas in the town's drugstore, Sherif bought a card at the counter and scrawled a quick note to Carolyn on the back, signalling the study's success. 'Dearest, the boys (members of the two groups) are having pops together in this drugstore in Heavener after their joint overnight camp. All going to Arkansas. Love M.' Then, scribbled in haste along the edge of the card, he added, 'See you tomorrow evening.'

Soon after returning to the road, they reached the state line and OJ pulled over. The boys piled out and gathered round a metal sign hanging off a plank, which read 'Arkansas Welcomes You', and one of the men snapped a picture. In the photo, a couple of the boys are looking upwards at a large boy, who has climbed to the top of the sign and is hanging upside down above their heads. I can't tell if he's a redhead, but it looks like the sort of thing Red would do. Only this time no one paid him much attention. The boys lined up, and one stood in the middle with his legs wide apart, with one foot in Oklahoma and the other in Arkansas, straddling the border of the familiar and the strange.

When they arrived at their destination — the small town of Waldron, in Arkansas — the boys jostled and laughed together over lunch in a diner, and in his notes Sherif wrote in happy capitals, 'EXPERIMENT ENDED AT THIS POINT.'

'We were delighted, just delighted with the outcome,' OJ said. 'Sherif challenged the idea that individuals are the problem or that they are inherently antagonistic. We created factions — we showed that by putting a group of normal eleven-year-old WASP boys in competition for highly desirable goals, you could mould them into factions. Then you could dissolve them again. It was an idealistic sort of thing for us, and we really felt we had a cure for our problems. We were fighting prejudice. But it was so busy that we didn't have time to stop and savour it — we just had to keep going.'

On the bus on the way back to Oklahoma City, the boys took turns in singing songs they'd performed during camp skits, with Bill Snipes accompanying them on the ukulele and leading the singing. As they got close, everyone 'rushed' to the front of the bus to join in singing Rodgers and Hammerstein's 'Oklahoma'. 'Everyone in both groups took part, all sitting or standing as close together as possible in the front end of the bus.

'The gaiety lasted during the last half hour of the trip; no one went back to the rear. A few boys exchanged addresses, and many told their closest companions that they would meet again',

*The Robbers Cave Experiment* concludes. It may well have been a triumphant moment for the men, but was it for the boys too? Did it mark the end of a fun and happy camp, or was their elation at least in part sheer relief that they were finally heading home and leaving the strange place that was Robbers Cave?

I understood now why Bill's memories of the camp veered between cheerful recall of happy times and the uneasy memories of the raids and the fighting. One of Bill's vivid memories was of a moment after they had left Robbers Cave for Arkansas. All twenty boys sat in the back of the truck, excited to be going to another state. The vehicle had a canvas tarp, and as they drove along the dirt road, dust blew back at them. Soon they were covered in it, their faces coated in the same brown mask. Opposite Bill, Red's pale, freckled skin and fiery thatch of hair was brown with dirt. 'I remember looking at that boy Red, the way his white eyelashes looked against the dust. And I remember thinking we all looked exactly the same,' Bill said. The differences between them had been obliterated. What's stayed with him is that final image, perhaps a vision of the camp as he had hoped it would be when he first arrived — a bunch of boys making new friends, who saw their similarities and not their differences, setting out together on a new adventure.

The 160-mile trip to Robbers Cave that OJ and Sherif made back in 1954 would have taken a lot longer than it does today, but still we left Oklahoma City early in the morning so we could make it there and back in a day.

Cherie, Bill's wife, drove us in her neat white Toyota hatchback. Bill had been cautious the first time I spoke to him about the experiment. Yes, he remembered the camp, he said. No, he didn't know it was an experiment. The words were familiar to me by then.

The traffic was light and we made good time, but after a while the double-laned highway narrowed to a single road and we passed gas stations, the occasional small town. Cherie and Bill pointed out the signs to the Okemah, where Woody Guthrie was born. The names of the towns are poetic: Shawnee, Keokuk Falls, Henryetta.

Robbers Cave in 1954 was Bill's first experience of summer camp, and he was looking forward to going back with me. He has a degree in sociology and psychology, and although he could not remember ever reading about the experiment during his studies, he was tickled pink to think he was part of something that's now in textbooks.

'I enjoyed that camp,' he told me. 'I had a great time.' Bill grinned at the memory. I was unsettled by this cheerful impression because it sat strangely with his and others' accounts of the fights and violence. Until Bill explained. 'My parents didn't have a lot of money,' he turned in his seat to tell me. The camp was a holiday for Bill that his parents would not normally be able to afford.

He hadn't been back this way in a long time, Bill said. He was clearly excited about this trip. Even his last visit down this way had been dramatic and frightening. In 1973, he and a bevy of officers from Oklahoma City got news there were riots at the state penitentiary and raced along the highway to get there. When they arrived, the place was on fire. It was chaos. The riots lasted for three days and left three dead and twenty-one injured. Twelve buildings burned to the ground. Bill didn't say much more, and I found it hard to imagine him raising a hand to anyone, let alone a baton or a gun.

We knew we were getting close to McAlester when we passed the state penitentiary. A sign by the side of the highway warned, 'Hitchhikers may be escaping prisoners', and I pondered the ambiguities of this statement all the way to McAlester. We parked at the small town of McAlester, the last stop before the turn-off, and had lunch in a diner with plastic tablecloths.

I guessed it was not often that Bill and Cherie got this far from Oklahoma City, and by this point I had caught the same mood of adventure they shared. I was grateful for their generosity in taking a total stranger on an outing like this, and on McAlester's main street, we took turns posing for photos beside a man-sized fish standing upright on the pavement outside the angler's shop. On the wall beside the shop, a poster advertised the annual Wild West Festival, which pays homage to the town's early beginnings, with a High Noon showdown, rope tricks, and a rodeo. Anyone looking at us would have thought we were ordinary tourists. But we weren't. For Bill, this was a chance to relive an early episode in his life. For me, it was an opportunity to visit the park for the first time with one of the boys as my guide.

When we stopped for petrol on our way out of McAlester, I spied a weekly newspaper in the rack called *OK Jailbirds*, which seemed to publish the mugshots and details of crimes for which people across the state had been arrested. I picked up a copy partly for its novelty value. Flipping through it, looking at the faces of alleged rapists, murderers, petty criminals, and paedophiles, I thought, *These are the kinds of people Bill has worked with all his adult life*. But he had none of the cynicism or seriousness I associated with someone who had done the job that long. He had a positively buoyant nature.

After we took the turn-off, the road slowly started to climb. The trees rose on either side of the road, and the flat grey landscape of the highway was replaced by a lush quiet forest. Finally, we turned into the entrance, Cherie parked the car, and we climbed out. Immediately we were hit by a wall of heat: it was like standing in front of an open oven door. Cherie and I wanted to stop in the shade and read the information boards at the park's entrance, to look at the tourist map and its faded black-and-white photos of famous outlaws, but Bill was impatient and hurried us on.

Trees shaded the road that snaked through the park, connecting the lower half, with the lake, to the upper half, with its cabins and cave. There was no one else around; we had the place to ourselves.

'There was nothing around it,' Bill had said on the drive, describing the cave to us. 'We climbed over these big granite boulders, twenty, maybe thirty feet in the air. They came out of the earth on a slant and you climbed up on them. You could climb real high.' For eleven-year-old Bill, this first week of the camp, exploring the park and climbing over and around the cave, had been a real adventure. He couldn't wait to get up there again.

We followed the signs to the cave. Almost immediately the path began to climb, but it passed under thick trees so I couldn't see where it ended. It was mid-afternoon, and we stopped often to catch our breath. Bill was the first to admit he was out of shape — unlike Cherie, who was fit and active. We took our time, stopping and resting and starting again. The trail was marked in fluorescent yellow, spots that in places had faded away or been obscured by bush. I expected to descend to find a cave: it seemed counterintuitive to be climbing upwards towards something I knew plunged deep into a mountain. It was an odd parallel of the feeling I had about Bill, who was treating the outing as a nostalgic trip to a place of his boyhood. How could he have enjoyed an experience that sounded so unhappy?

When we got to the top, it was exactly as Bill had described it. The trees cleared and we came to a flat plate of rock that jutted out above a valley, surrounded on all sides by forest. Up here, the road we'd driven in on was invisible, and on the rock we could look out across a wooded valley and to the Ouachita Mountains. Insects hummed in the otherwise silent wood as we wiped the sweat from our faces. We seemed to be the only people for miles around.

Bill was disappointed that we couldn't get closer to the cave, but barriers had been erected around it to prevent the public from entering. Looking at the cave mouth from where we stood on a wooden walkway, I could see how the light stopped abruptly just inside, swallowing the sun in a thick darkness.

'We used to climb down to there.' He pointed to one side of the cave's entrance. 'It was really steep, and we'd slide down in there. You didn't know how far you were going to go. It was very dark ... But it was fun,' Bill protested quickly when he saw Cherie and I exchanging glances. 'We had stakeouts. We found some great hiding places. I had a great time.'

Bill's face was shiny with sweat and he was still puffing from the climb. We decided to take a break before we began the descent.

I climbed back up on top of the rock, above the cave. I could see why ten- and eleven-year-old boys would have loved this place. Standing on the slab of sandstone jutting out across the valley, I could imagine that Jesse James stood in the same spot, keeping a lookout, back in the days when there were 'cowboys' and 'Indians'. There was a kind of spell over the place. To city boys like Bill, it must have seemed magical.

I could hear Bill talking loudly below me. 'Guess what I'm doing?' He sat on a rock, his face pink, a cell phone to his ear. 'I'm climbing a mountain! I'm finally gonna get into shape.' He chuckled. One of his daughters, clearly incredulous on the other end of the line, was making him laugh.

The stone-and-wood cabins looked exactly as they did in the 1954 photos. The camp caretaker unlocked the door of one and let us have a look around inside. The park is busy in the cooler months, when the snakes are not active and the weather is bearable rather than blistering, and she told us they often rent the cabins to families, former school classmates, or members of church groups who camped here together in their younger days and now return for reunions. But the cabin I was in smelled as if it hadn't been used in a while; it was musty and crowded, with old wooden bunk beds harbouring yellowing mattresses. Bill swore it was the one the Rattlers had slept in, and outside he headed off through the trees, over a creek, to the Eagles' cabin.

Bill made a slow circle around it. 'I remember when I came through that window, Davey Munroe's eyes were out on stalks. He was not happy, he was real kind of upset.' Bill looked troubled at the memory. He stopped, and Cherie handed him a bottle of water. 'You know I played football with Davey Munroe at college?' He unscrewed the bottle and took a sip.

'You did?' It was the first instance I'd encountered of any of the boys crossing paths in later life, and I was excited.

Bill shook his head. 'It's the strangest thing. We talked a lot about other things, but we never mentioned this camp. You'd think we would, wouldn't you?'

'Maybe he didn't want to remember you falling on top of him,' Cherie joked.

But Bill went on as if she hadn't spoken. 'I mean, this is where we met, and every time I saw him I would think about this camp. I just don't know why we never got together and talked about it.' He looked at me hopefully. 'Boy, I would love to meet up with those fellas again.'

Bill returned to this topic like a tongue at a sore tooth, and we talked about it a number of times over the next couple of days, but he wasn't able to explain it any better than this first time. 'I have no idea why Davey and I never talked about it,' he said, shaking his head sorrowfully. 'Or why I never stayed in touch with these guys!' This is not at all characteristic for Bill. This is the man who meets for coffee every Thursday with his friends from elementary school to reminisce and catch up on gossip. Bill is clearly proud of these longstanding friendships. Which is why he's so bothered by Robbers Cave.

And yet Bill recalls so much detail — not just the description of the cave, but the first week exploring and playing, the bully Red, the night raids, and finally the happy end stage, where the boys all seemed to get along, and Bill led the singalongs on the way back home, playing his ukulele.

I wouldn't be able to put Bill together with Davey Munroe, but I had managed to track down another boy from Bill's Rattler group. Maybe meeting and talking would put whatever it was that was bothering Bill to rest.

Smut Smith had been on Google, reading about the experiments, ever since I first made contact. He was excited to see himself online in a picture at the head of a group playing tug of war. 'I said, "That's me!" I called my wife over to look and she said, "No doubt about it, that's you."'

Smut has a strong Oklahoma accent and lives in Edmond, not far from Bill Snipes, although they haven't seen each other since 1954. 'We really had fun,' he told me. 'We swam a lot. We had our own pond, and even though there were snakes in the water, we'd make a lot of noise and splash around a lot so they wouldn't come near.'

What were the experimenters trying to find out, he wanted to know. I gave him a brief rundown and he was silent for a while. Then he said slowly, 'I remember the rivalry. I remember we tried real hard to win in all these games. But it seemed they had better athletes, and they dominated us. We didn't like that. I remember we ganged up against them and plotted out strategies, but I can't remember the specifics.

'I remember I made some good friends, some real buddies. But I remember one guy, he was a real bully. He was bigger than anyone else and was always pushing people around.' Smut veered between recounting the conflict with Red and with the other group. But both he and Bill returned to Red a number of times.

The other thing he remembered was the relief when it was over. 'We were there for three weeks, and I didn't realise how homesick I was until I saw my folks. I remember crying when I saw them again.'

He was voluble on the phone, excited that he was in what someone had told him was 'a world-famous experiment'. I would have thought he would be keen to meet one of the other boys, but although he agreed, I sensed some reluctance.

A week later, Bill and I met Smut at the Cracker Barrel, a chain restaurant that boasts traditional home cooking, located halfway between Oklahoma City and Edmond. I took a photo of them together before we went inside. Smut, tall and straight and fair-haired, in a polo shirt, smiled in a restrained kind of way. Beside him, Bill, short and stout, in a loud Hawaiian shirt, grinned widely, but I was surprised by how nervous he seemed. Inside, waitresses in long gingham aprons took our orders. The conversation proceeded in fits and starts. Both men had brought their wives. Bill asked Smut what line of work he was in, and they exchanged details of college and careers. I realised what a thin thread tied these two men together. Smut was reserved, cautious. Bill too seemed wary. I was fiddling with the straw in my drink, wondering why Smut had agreed to meet us if there was so little to talk about, when Bill mentioned the frog. Smut's head came up and his glasses flashed.

'You remember that canoe?' Bill said, grinning.

'Pfft.' Smut was dismissive. 'Six or eight of us in it.'

'That counsellor with the horn-rimmed glasses —'

'He was with us a lot,' Smut said. 'I remember he was trim.'

'Around my dad's age,' Bill said.

'And that terrible noise.' Smut grimaced, laughing.

'I never saw anything like it. That frog was the size of a saucer!'

'We floated right up to it.'

'Each snake with a hind leg in its mouth, and rest of snake

strung back out on the water —'

'That frog was trying to get loose. Awful noise, wasn't it?'

'And that counsellor drew a pistol.'

Both men laughed.

'It was exciting,' Bill said.

'Sure was,' replied Smut.

They were silent for a minute.

'You think he shot that frog,' Bill said, 'or did he kill the snakes?'

'Hard to tell,' Smut said.

They mulled it over a while, jiggling straws in their drinks. Had the man saved the frog or killed it? Was he villain or hero, friend or foe?

'They studied us,' Smut said. 'What did they determine from their studies? What were they trying to find out?'

It dawned on me why they might have been nervous about meeting. They sensed that the experiment had been a kind of test. But whether they had they passed or failed, and what the men had learned about them: that they didn't know. When they looked to me to explain the experiment, I reiterated what they already knew, that it had been about how you could bring groups to war, and then bring them round to peace again. What I left out was that Sherif later defined it as a kind of moral test that the boys had failed.

I felt sure this doubt about the experiment's significance was at the heart of Bill's uneasiness. Perhaps he hoped in meeting one of the other boys to find the answer, but it seemed to me that Smut was cautious too and might have been bothered by the same questions.

Despite the final reconciliatory stage, where the boys merged into a single harmonious group, the Robbers Cave boys are famous for their supposed transformation, in Sherif's words, from 'cream of the crop' to 'disturbed, vicious … youngsters'. And if you read Sherif's description of the experiment and the version repeated in psychology textbooks, you get the same story.

But how was it that the boys at Robbers Cave turned on one another in this way? They were no less 'normal' than the boys from New York State. They were the same age, and while they were poorer and from down south, it seemed unlikely that socioeconomic background or geography explained the change.

The hypothesis and design in 1954 were different: in Oklahoma, Sherif and his team had dispensed with any attempt to convert friends into enemies. The two groups had no chance to make friends before the competition started — even though the boys had tried, they'd been blocked from developing intergroup relationships. Jack White's refusal to invite the other group into the birthday party got the tournament off to a hostile start. Unlike the year before, the norms of good sportsmanship soon dissolved, as at Robbers Cave both groups typecast their opponents as cheaters. The Eagles sat down during the tug of war, which was against the rules. The Rattlers trashed the Eagles' cabin the night before cabin inspection to gain an edge in the competition.

Was it also the behaviour of the researchers that was different this time round? I revisited the staff instructions and noticed Kelman's original directives had been revised for the 1954 study and contained new ordinances. Once boys decided on a 'line of action', the book noted, staff could 'give them help to carry it out' and 'give advice'. In contrast to the 1953 study, where staff were instructed to stay at arm's length and not 'influence' the campers, this time staff could take a more active role. The boys were likely 'to turn to you, as adults, for approval or sanction', and as long as the boys' actions 'do not run counter to the criteria for a given stage', staff were allowed to give the OK for them to go ahead. But exactly what behaviour the men could permit and encourage, and what they would forbid, the instructions didn't say. So they had more leeway to shape the group's behaviour in line with the experiment's hypotheses and to move things along. Remembering how much pressure OJ said they were under, I wasn't really surprised. There'd been a few instances when I was going through the materials where it seemed to me that staff were clearly crossing a line, but I could see now that they were within the rubbery boundaries of the directions they'd been given.

A year earlier, Jim Carper had made a habit of ensuring the whole group voted and came to a unanimous decision on any action they wanted to take, but there's no mention of this process at Robbers Cave. So just exactly how many boys had to suggest flag burning before staff 'gave a hand'? I thought of the matches that mysteriously appeared when the Eagles boys stood around the Rattlers' flag, and how, if following these instructions, it would have been acceptable for the men to help them set the flag alight.

So where exactly did the line between staff intervention begin and end? Surely the men's involvement in and approval of retaliation and vandalism had influenced the boys' behaviour? Bob Hood not only didn't reprimand his Rattler group for trashing property but also accompanied them on the night raid and took photos. The men blocked behaviour that might run counter to their hypotheses too: after the first raid, the Eagles suggested a peaceful resolution to Jack White, asking for the Rattler group to be disqualified from the tournament, but White suggested they sabotage the Rattlers to even the score.

How much of the groups' 'initiative' originated with the boys and how much with the men? Whose idea was the night raid, the smearing of their faces with soot 'commando style'? How long did Bob Hood and Jack White use the excuse of a bet with each other to motivate their groups to win? The observation notes weren't nearly as informative as a source for me this time round. They were much shorter, and fulfilled a different function too. Instead of trying not to pay selective attention, as Herbert Kelman had urged in the earlier study, this time the men were urged to do the opposite: 'The ongoing activities will present the possibility of an infinite number of events that could be observed and recorded. Therefore, please have the hypotheses for the given stage focal in your mind so that observations will not be hodgepodge but will be relevant to the hypotheses in question …'

One of Sherif's major predictions, and the one that has made Robbers Cave famous, is the apparently spontaneous mistrust and hostility that erupted between the groups during the tournament, and the apparent inevitability of the fighting that broke out when the Eagles group won the prize. But rather than competition causing conflict at Robbers Cave, it was the intervention of the men setting the groups against one another that added fuel to the fire. The year before, Kelman objected to these 'frustration situations' organised by the staff to increase hostility between the boys, calling them 'a serious violation' of the experimental design. In the 1953 study, the staff, playing agent provocateurs, created incidents so that one group of boys would mistakenly attribute it as an attack from the other boys. Their interventions had been relatively benign — mixing up clothing, cutting a flagpole rope — but neither group took the bait. At Robbers Cave, Sherif and his team upped the ante. But the hostility between the boys couldn't be explained as a byproduct of the competition; it was because they felt under attack. They fought one another not because one group won the tournament but because they had been violated, their flags burned, their huts raided, their prizes stolen.

It seemed to me that what happened at Robbers Cave wasn't a test of a theory so much as a choreographed enactment, with the boys as the unwitting actors in someone else's script.

The men had encouraged hostility and fighting. Now I thought I could see why OJ would so vividly remember the 'kangaroo court' and why it was not included in the final report on the experiment. I thought I understood why some of the Rattler boys, such as Bill and Hollis, had kept the men away when they wanted to take Red in hand. The men had never stepped in to protect boys in either group from Red's bullying and aggression. If anything, they rewarded him for it. It seemed no coincidence to me that the afternoon these boys banished Red and stripped him of his power was the same in which the two groups made tentative steps towards friendship. Sherif might have put it down to the power of a superordinate goal, but I read it as the boys restoring rules that the men had broken.

When I tried to pin OJ down on just how much of a role the men played in generating friction between boys, he said, 'I might be biased, but I think the Robbers Cave study was clean. We introduced manipulations and then we stepped aside.' But this bothered me. How could they both manipulate and step aside?

'What we told them was they'd have to do things safely. We made it very clear that they could essentially do what they wanted as long as no one got hurt.'

I thought how frightening that idea must have been to some of the children. They were in a remote rural wilderness with men who intimated there were few holds barred and where the values of home and church and school had no force. Respect and fairness were discouraged. Cursing, bullying, cheating, and fighting were rewarded. And in the notes of *The Robbers Cave Experiment*, the boys' misery and resentment spills across the pages.

There was a kind of naivety at work here in OJ's response that surprised me. In an experiment about group influence and inequities in power, the men seemed blind to their own role as a powerful group in the camp. On one hand, they acknowledged how they engineered events and set up misinformation so that one group would get angry with the other and retaliate. But they couldn't, or wouldn't, see that their non-intervention was in itself an action that sent a powerful message to the children. The fact that they did nothing to prevent or put a stop to the name-calling and the cursing, the food throwing, the vandalism, and the raids communicated their approval and encouragement. How could they state that they had manipulated the boys' interactions at the same time as arguing that the boys' behaviour was 'natural'? I was surprised, too, that someone like OJ could be so insightful in some ways but not have reflected on this.

Sherif used the term 'hypothesis' in his descriptions of what he planned for this research, but I was struck in reading his research proposals and outlines that there was nothing tentative about his predictions. The book that resulted from the 1954 Robbers Cave study describes a surprisingly neat scenario that proceeds smoothly through each stage, in vivid contrast to the Middle Grove study the year before:

> … in testing our main hypotheses, we supplemented the observational method with sociometric and laboratory-like methods. One distinctive feature of this study was introducing, at choice points, laboratory-like techniques to assess emerging attitudes through indirect, yet precise indices. Such laboratory-like assessment of attitudes is based on the finding that under relevant conditions, simple judgments or perceptions reflect major concerns, attitudes and other motives of man.

Despite the often dry and scientific language of passages like this in their book and an emphasis on careful testing, Sherif's study was not an experiment in the sense of an investigation, a gathering of evidence or proof. Sherif's theory, already elaborated and developed, was the road map. The job of the research staff was to follow it.

So Sherif had a clear idea from the outset of what would happen, I asked OJ. 'Oh, he had a definite script in mind, all right. His mind was not open. It was most definitely made up. And it was me who looked after the logistics, who made it happen.' I was taken aback by how matter of fact and unapologetic he was about it. As if there was no question that this was the job of the research team, to deliver on the scenario they had already envisaged, like stagehands to a director.

I remembered the first time I came across a groundbreaking paper on the ethics of psychological research, when I was researching my book about Stanley Milgram. The paper was written by a professor called Herbert Kelman, and titled 'Human Use of Human Subjects: the problem of deception in social psychological experiments'. Published in 1967, it was an in-depth discussion of what psychologists should and shouldn't do in the name of social psychological research with human subjects. Before then, published discussion of the ethics of deceiving people about the purposes of an experiment was rare, and Kelman's paper was a watershed moment.

I read the paper in preparation for my interview with him in Boston in 2013. Part of the text reads, 'There is something disturbing about the idea of relying on massive deception as the basis for developing a field of inquiry' as it 'establishes the reputation of psychologists as people who cannot be believed'. So when I met Herb Kelman, I was eager to find out if this influential paper was inspired by his experience working on Sherif's summer-camp experiment.

'I'd never made that connection,' he said with surprise. 'If I was still in therapy now, that's something I'd take to my therapist. I've written a lot about ethics of experimentation, but this study never figured in my thinking — you've just opened up a question for me. I really do think now that you raise it that there are serious ethical questions about it — there's a difference between deceiving them in the course of a one-hour experiment and deceiving them about a three-week experience. I think you're absolutely right. I'm sorry I'm not in analysis right now; I'd take it up with my analyst: *How come this hasn't figured?*'

Perhaps it didn't matter that he hadn't made the connection consciously; his paper fanned a public debate about ethics that would change the way psychological research was conducted. After talking to OJ Harvey about both experiments it seemed to me that he and Sherif did not consider the ethical implications of their studies. They considered the boys' physical safety — the risk of snakes, bug bites, black eyes — at the same time that the observation notes showed that there was evidence of the boys' emotional turmoil, the boys' crying, praying, and bullying. How could they notice such behaviour and not worry about its effects?

There was a kind of double vision at work here. To his professional peers, Sherif described the study as about nations, states, hostility, and conflict. But to anyone concerned about the ethics of the research, Sherif argued that the hostility between the boys was little more than schoolboy rivalry:

> The competitive situation consists of games which boys enjoy playing. The enthusiasm in the activity and in opposition to the out-group in our work has been by no means more, and probably less, than can be observed on any street corner or school between rival cliques or school teams composed of normal healthy boys. For example, here in Norman the football game between McKinley and Lincoln *gradeschools* engenders greater excitement than the situations I specified in my experimental outline …

This denial of the boys' experience extended beyond the study. OJ told me that Bob Hood regularly talked about the experiment in his teaching at the University of Oklahoma. After one of his psychology classes in the 1960s, a student approached Hood to say that he was one of the subjects. But Hood brushed him off and told him he was mistaken.

In the lobby of the National Cowboy & Western Heritage Museum in Oklahoma City is a huge white sculpture of a Native American slumped over his horse, his lance hanging uselessly from his hand. Both man and horse look defeated and on the brink of death. The plaque reads 'The End of the Trail', and I was contemplating why this sculpture of a dying American Indian was given such a prominent place in the museum when my phone rang.

I'd left a message for him three weeks earlier, but Dwayne hadn't called me back. All I knew about him was he worked for a company that makes medical devices and that he sounded businesslike and rather brisk. I was pleased to hear from him, and the last thing I was going to do was tell him that now wasn't the best time. I took my phone outside and sat on a chair at the edge of the museum café, away from a table of grey-haired tourists in glowing white sneakers. Since we last spoke, he had started reading the book about the experiment and had waited till he finished it before he rang me back.

I felt a clutch of anxiety to hear that Dwayne had read the book, and I realised how protective I was beginning to feel about these adult men. I'd worried about what it might feel like to recognise yourself in *The Robbers Cave Experiments*, how it could be upsetting to learn you could say or do things that you didn't know you had in you. It was one of the reasons I had been reluctant to recommend the book to those men I spoke to. It was written to emphasise the way the staff simply stood back and let things take their course, and in reading it, one could easily believe that the boys weren't manipulated.

But not Dwayne. 'They were out to prove a point, weren't they? It didn't seem very scientific to me. I mean, what did they prove — that you can set things up so people will argue and fight? That's news?' There was an edge of impatience in his voice.

'You encouraged the others in your group to pray before games,' I said.

'I bet I did,' he said dryly.

'You're not still —?'

'No!' he laughed. 'That was my parents' influence. I grew up going to church, but I gave it up in my teens. I was a "troubled teen".' He put emphasis on the phrase. 'I gave my parents a terrible time — stopped going to church, got in with a gang. For a while I even dropped out of school.' As if he'd read my mind, he said, 'I'd love to blame it on this camp, and who knows, you know, maybe all that fighting they had us involved in did something. I don't really believe that, but it's always been a bit of a mystery to me: where did I get the idea — this mild little goofy kid — where did I get the idea just a couple of years later that it was okay to solve problems with my fists? Don't think I haven't thought of it since all this came up.'

He said something else, but there was a burst of laughter from the tourist table and I missed it. 'Sorry, could you repeat that?'

'They thought they were doing the right thing, my folks,' he said, clearing his throat. 'They were eager to send me. They thought it would be good for me, you know.' There was real sorrow in his voice. 'I guess I learned to stand up for myself, that not everybody plays fair.'

Chairs scraped. The tourists were leaving.

'How did they get away with it?' Dwayne wanted to know.

After I got off the phone, I wandered around the museum, among tributes to rodeo stars and stunt riders, but I wasn't concentrating. I was thinking about Dwayne with a vague sense of uneasiness, as if I'd done something to be ashamed of. Perhaps in looking for these lost boys I had been just as blind to my own motivations as Sherif appears to have been, and I was blithely stirring up old hurts. It seemed that as far Sherif and OJ were concerned, as long as no boy was physically injured, everything was fine. They ignored the boys' emotional states almost entirely, except insofar as they had a direct bearing on the success of the experiments.

But, from that early, casual afternoon in the archives, over time I had become just as single-minded in my quest to find the boys and establish some link between their experiences at the camp and their later life. I'd set out to discover how the boys involved were affected then and later, and how the experience had changed them. I had a theory, a hypothesis, that perhaps they carried some hidden legacy from the experiment that shaped who they were today. I'd thought of them as lost boys in the sense that they had a missing part of their childhood that as a researcher I could restore to them. But without realising they didn't know they'd been experimented on, I had blundered in, not always duly considering that the information I was delivering might be intrusive or unwelcome. Or that the news they were part of an experiment could disrupt their sense of themselves and their life story in ways that took time to unfold. Absorbing this information and making sense of it was a process.

The next morning I prowled my hotel room, checking emails, flitting between internet tabs, waiting for something that I couldn't name. I went for a walk through Oklahoma City's old warehouse district, down along the canal, hoping that physical exertion would settle my restlessness. But the further I got — past the ballpark and Mickey Mantle Plaza of Bricktown; along the shaded canal, with its early morning emptiness — the more convinced I became that I'd missed something. I stopped and watched a man step onto a canal boat, holding the hand of a small boy who struggled and tried to pull away, clearly wanting to get on the boat by himself. I thought of the boys, now men, I had interviewed and the painful new knowledge that they had been deceived, and I thought of the man who had deceived them. Muzafer Sherif inspired and attracted a dedicated team of researchers, yet none of his male colleagues I'd spoken to so far remembered his charm or garrulousness as much as his demanding temperament and drive. I realised that despite all my research, Muzafer Sherif remained a troubling and enigmatic figure to me.

Was he just so driven by ambition or idealism that he was blind to anything else? But he undermined his own study on many occasions with his self-destructive behaviour, his inability to manage staff, his drinking. Could there perhaps be a psychological scar in his own past that could explain his apparent lack of compassion for the boys? Who knew? He spent his formative years in Turkey, I knew that, but apart from this rudimentary detail, for all the hours I had spent examining the files and the documents, reading this man's letters and notes, there was still a mystery at the centre of the story, and that mystery was Sherif. He had heart and soul invested in this group research, but I wasn't any closer to understanding why. I wouldn't understand the story of the Robbers Cave until I came closer to understanding the mystery of the man who orchestrated it.

# Part Three

# 10
# Empire

In the summer of 1914, a group of village boys kicked a ball along the road, shouting and sending up puffs of dust. They moved in a pack, chasing a donkey and cart along the road, jostling at the village well to take long drinks of water, slowing down only when the sun began to wane.

Muzafer Sherif was one of them. As usual, he'd discarded the leather sandals that marked him out as wealthier than the other boys, the sons of his father's tenants. When the camel caravans passed through Ödemiş — an Anatolian market town in the Aegean region — on their way from Egypt, the drivers stopped at their favourite tea house. It would have been a typical Turkish establishment, with a large tree outside that kept the interior of the wooden building dim and cool even during the hottest part of the day.

While the men sat inside with glasses of tea, the boys, led by Muzafer, sneaked behind the building where the camels were resting and restrung their ropes. When the drivers returned and the camels stood again, they were not tied in single file but every which way. It was chaos. The men shouted, and a few ran from the square, looking for the culprits. But the boys were gone, racing through the narrow streets, and I pictured them gasping with laughter and exhilaration.

The next time, the drivers left one man behind to hide and keep watch while they had their tea. Muzafer and the other boys were in the middle of retying the camels when the drivers — tough men used to fighting off bandits and thieves — charged towards them, yelling and waving their knives. In the shouting and confusion, the boys scattered — all except for Muzafer, who, caught between two camels grunting and kicking in fear, was knocked to the ground. He lay unconscious in the dust, and when he came to, he was surrounded by the furious camel drivers. I don't know how he got away; perhaps one of the other boys ran to Muzafer's home and got his father, who was also the town's mayor, to hurry to his son's rescue.

Fifty years later in America, Sherif had a brain scan following a stroke, and his doctors were mystified to see evidence in his X-rays of serious head trauma. It was then that his wife, Carolyn, recalled the story about him being kicked by a camel. Sue Sherif, their daughter, told me this anecdote. Sue wasn't able to tell me a lot more about her father's early years in Turkey because he didn't like to talk about it. But she gave me what she could: snippets of his childhood, the name of the village where he was born. Muzafer was born in the summer of 1905 in a mountain village above Ödemiş called Bozdağ after the mountain range's highest peak.

The taxi driver dropped me off on the edge of Bozdağ, at the one hotel, which was closed. Brown leaves lay on the bottom of the swimming pool in inches of muddy-looking water. Except for a tractor that chugged down the centre of the street, trailing a cloud of diesel fumes, the place was empty. Under the shade of the awning, a small strip of village shops displayed their wares. Straw brooms poked out of a bucket; earthenware pots were stacked on a table covered with a chequered plastic cloth.

On every side, the mountains rose above the red-tiled roofs. It was September 2015, early autumn, and between tourist seasons. The summer rush was over, although the shop in the main street still had brightly coloured beach balls for sale, hanging in a string bag, for swimming excursions to nearby Lake Gölcük. It was a warm afternoon, and the winter season — when the village turns into a ski resort for locals — seemed a long way off.

I had no more than a few words of Turkish, and I had worried about how I'd get by in the country, especially further away from Istanbul. But I'd negotiated a fair price with Osman, the taxi driver, and he kept up a steady stream of conversation in Turkish as he drove across the valley and up into the mountains. He pointed out the window at the countryside as he drove, the blue eye amulet swinging wildly from the rear-view mirror, and I smiled and nodded from the back seat as if I understood.

Strings of chillis hung in pretty red loops against a stone wall; latticed wooden balconies threw lacy shadows. I felt optimistic, convinced suddenly that I would do more than just get by here. I had done as much reading and preparation as I could before I arrived, and even though I'd found next to nothing in English about Bozdağ, I knew enough to see that this sloping street, lined with ancient stone houses and terracotta roofs, would have looked exactly the same when Sherif was a boy.

Under the blue awning, a small group of men looked up from their game of backgammon as I drew close. One of them stood up and rushed over to talk. 'Where are you from?' he called in raspy smoker's voice as he approached. He had a cloth cap pushed back jauntily on his head, and a thick white moustache.

'Australia,' I called back.

He nodded confidently. 'Sydney? Perth? Darwin? I've been to all of them,' he said proudly. 'I was a sailor. I went all over the world.' He dragged on a cigarette. 'You like Turkey?'

I nodded enthusiastically.

'What are you doing in Bozdağ?' he said, tipping his head to one side.

I hesitated. Was he old enough to have known Muzafer Sherif or his family? 'I heard it was a pretty place,' I answered eventually.

He shrugged and looked up and down the empty street, then back at me.

'How long have you lived here?' I asked, trying to calculate his age. But the sun had turned his skin leathery and brown, and it was hard to guess.

'No, no,' he said, flapping the hand with the cigarette in it at me, as if the idea was preposterous. 'I am visiting my cousin.' He gestured towards the players at the table. One of the men called something to him and he muttered back. He threw away his cigarette. 'Have a safe journey!' he said, and strolled back to the backgammon table and took his seat again.

I'd decided to keep the fact that I was a writer to myself. Otherwise it meant I had to check in at the police station of every village and town I visited. But I could see now the problem with that plan. It was difficult to have genuine conversations with people if I lied about what I was doing here.

I wandered down the street and stopped in the centre of the village, outside the mosque, and squinted up at a statue of what looked like a soldier in baggy shorts and black boots, holding a rifle and gazing off at the mountains. Beside him, there was a large framed black-and-white photo of Mustafa Kemal, bareheaded and in what looked like a tuxedo, who peered seriously out at me from under the splash of red, the Turkish flag. I was trying to figure out the relationship between the two of them when I heard the toot of a horn, and Osman's taxi rolled to a stop beside me.

Osman was impatient with my desire to walk through the small village. Clearly, for him, the attraction of the place was the mountain itself. He followed the road up a steeply wooded mountainside, where he passed through large iron gates and then parked in a clearing beside some dilapidated picnic tables and a small tea house set up under the trees.

A cool breeze came off the mountain peak above, but it was invisible behind the forest of oak and pine trees. Mount Bozdağ was capped with snow year round, and back when Sherif was a boy, enterprising locals had climbed the peak and brought back pieces of ice wrapped in felt that they sold for refrigeration. Two miles above sea level, the 2,000-metre-tall chain of mountains had been a magnet in summer for wealthy Levantine families wanting to escape the heat of the plains. They arrived with their servants and camels loaded with tents and furniture, food and utensils, for a season of hiking, blackberry picking, picnics in the forest, and boating on Lake Gölcük.

The mountains were also a great hiding place for local brigands or outlaws. The sultan's authority only extended so far.

When Sherif was young, groups of armed bandits, or Zeybeks, lived in these mountains, beyond the reach of law and order. They thundered down the mountains on horseback to rob traders, kidnap for ransom, or demand protection money from those in passing caravans. It was a great place to hide. One local brigand named Charkirge had a price on his head for kidnapping and holding wealthy hostages for ransom. The summer Muzafer was born, the Levantine community holidaying in the mountain resort were panicked by rumours that Charkirge, frustrated by the birth of yet another daughter, was planning with his followers to kidnap and keep a baby boy. The rumours panicked a missionary family who were holidaying nearby, and in particular the new mother, who was recuperating in the mountains after the birth of her son. In her memoirs, she recalls how she appeased Charkirge by taking gifts to his wife and hosting a lavish picnic on the mountain in their honour.

The owner of the tea house insisted I climb up into a kind of treehouse, furnished with cushions and a low table, and brought me tea in a small glass rimmed with gold. He and Osman settled at a table in the shade below, and the smell of their cigarettes drifted on the breeze. It was quiet except for the rustle of the wind and the low gurgle of a nearby spring.

Yet soon the tea house owner was back, gesturing at me to follow him. Past the picnic tables, he pointed at an ancient chestnut tree, which, a sign beside it told me, was 20 metres high and 3 metres in diameter. I ran my hand over the trunk, feeling for carved initials. The tea-house owner showed me a pipe that gushed mountain water, encouraging me to cup my hands and take a drink of water so cold it made my teeth ache. As a boy,

Muzafer would have drunk from this same stream, thirsty from games of hide and seek when he was little, after hikes or horserides up the mountain with his brothers as he got older. In spring, the floor of the forest was carpeted with yellow, white, and blue crocuses. The trees threw a welcome shade, where he would have been able to sit on a bed of soft pine needles when he got tired. How long was his childhood peaceful, I wondered; at what age did he become conscious of the violence that was never far away?

The year Sherif was born, there was widespread revolt supported by the Committee of Union and Progress, or the Young Turks, against the sultan's unjust taxes. Three years later, the Young Turks led a revolution, restored the Ottoman constitution of 1879, introduced a multi-party political system, and ended Sultan Abdülhamid's thirty-year autocratic rule, replacing him with his younger brother, who was little more than a figurehead. But the euphoria — the Muslims, Christians, and Jews celebrating and dancing in the streets — did not last long.

Ödemiş felt like a metropolis after Bozdağ. Due to the upcoming election, Recep Tayyip Erdoğan's face, ubiquitous as it had been in Istanbul, loomed from the side of buildings or from billboards by the highway. The wide tree-lined main street, with its imposing stone buildings built in the early years of the Republic, was lined with restaurants and grocery stores. Cars roared, motorcycles buzzed. Osman dropped me a block from my hotel, and I paused at a café with a large shaded terrace and thought about having a cold beer. It looked like the sort of place that would serve alcohol, but perhaps secularism was losing its grip here in this Kemalist stronghold too. No one else seemed to be drinking alcohol, so I returned to my hotel.

One of Muzafer's brothers, a lawyer and prison reformer and, later, chief magistrate and mayor of the town, had established a museum of his own on the edge of Ödemiş, housing a collection of more than 16,000 artefacts from the area that he acquired through his lifetime. But the museum looked like a building site; a security guard shook his head at me through the locked iron gates. A harassed-looking woman came out to tell me the place was undergoing renovation and was closed.

The clerk at my hotel, a modern high-rise that seemed set up for Turkish businessmen, handed over my key with a curt nod. I had the feeling he was wondering what I was doing in Ödemiş. It had been a frequent question from shopkeepers since I arrived three days ago. The man behind the counter at the local grocery store had said bluntly, 'There's nothing for tourists here. You should be at Ephesus.' It came across as unfriendly, unwelcoming. It was not a comfortable feeling, and I was beginning — irrationally, I knew — to dislike the place.

Muzafer Sherif's father, a member of an emerging Muslim bourgeoisie, moved his family from Bozdağ down to the plain to Ödemiş when Muzafer was small. Here he acted as a broker between local farmers and British traders (the British had brought the railway line in 1884) and organised the sale and transport of tobacco and cotton to nearby Smyrna for export abroad. Muzafer's grandparents lived in the medieval village of Birgi, famous for its silk production and its Islamic scholars, where his grandfather was a teacher. In the town of Ödemiş, Muzafer's family was relatively affluent, with servants to do the housework and tenants to work the land. Later, Muzafer's oldest brother would serve as town mayor.

It was the Republican People's Party (CHP) mayor of Ödemiş, with his passion for the town's history, who was one of the main organisers of a symposium in honour of Muzafer Sherif in 2013. But by then no one in Turkey remembered what Sherif looked like, and the organisers mistakenly chose a photograph of social psychologist Solomon Asch for the printed program and projected Asch's image onto the wall during the ceremonies. In the audience, few other than Sherif's daughters, visiting from America, noticed the error. 'That's not Daddy!' they whispered to each other, debating whether to say something. In the end, they decided it didn't matter and stayed quiet. Sherif would have been enraged, especially given that he credited himself with inspiring Asch's interest in social psychology and later came to regard him as a rival. After the symposium, the mayor of Ödemiş snipped a ribbon at the entrance of a small street in the old part of town and announced that it was now named Muzafer Sherif Street.

But the clerk at the front desk of my hotel couldn't help me find Muzafer Sherif Street. Like me, he checked Google Maps, so I knew he wasn't going to have any better luck. He tried a few spelling variations, then looked up and shook his head. He typed something into his laptop and swivelled the screen around for me to see. Ödemiş had a city museum that he was urging me to visit.

The local cultural museum was a two-storey eighteenth-century wooden inn with the rooms arranged around a central courtyard. The fourteen rooms on the upper and lower floors had room after room dedicated to local culture, leading citizens, and their business ventures. In the women's quarters, in the dowry room, a Singer sewing machine stood ready, and mannequins modelled ornately embroidered wedding gowns. In the male rooms, life-size dioramas showed local businessmen going about their daily life in rooms decorated as their shops — the local clockmaker squinted down through his eyeglass into the innards of a watch, the barber sharpened a razor, the shoemaker fashioned leather.

Tuğba the archivist, a slight woman with a mass of dark curls and large glasses who was clearly very proud of the museum, took me upstairs and showed me a small glass cabinet in the corner of a hallway. Tuğba's English was rusty and my Turkish was nil, but I understood from her that this display was the result of the commemorative event for Sherif held back in 2013. On display inside were four of Sherif's books, which looked measly in comparison to the dioramas celebrating the town's other residents.

I told Tuğba I was looking for the street in Ödemiş that was named after Sherif, and she consulted a book and came back with a slip of paper in her hand, offering to call me a taxi to take me there.

On the street she handed the driver a piece of paper and gave him directions. He looked at the address, shot her questions, and then opened the door doubtfully. 'He will take you,' Tuğba said confidently.

But soon we were lost. The driver stopped at a shop to ask directions, then at a shoemaker's stand, under a striped canvas awning on the footpath. Finally he stopped at what looked like a car park, where a sad-faced man got up from his desk and came out to inspect me through the window of the cab, then pointed across the street.

It was a long street, lined with low stone houses hidden behind high walls. The taxi driver stopped at a pink-washed wall with a blue door, the branches of an olive tree spilling down over it. He pointed at the house, then down at the paper Tuğba had given him. I walked to the corner, looking for the sign, but there was none. No wonder no one knew where the street was. Further along, the old houses petered out, and it widened into a quiet roadway that stretched away to the distant mountains.

It definitely wasn't the same place that I'd seen in the photos of the opening ceremony. In the pictures, the mayor was lined up with a row of dignitaries at the entrance of a small and crowded street beside the blue-and-white sign bearing Sherif's name. But it didn't matter. The house in this unmarked road must have been his childhood home. Whether it was a language miscommunication, or perhaps because Tuğba knew that the street named in his honour was gone, as I later found, she had pointed me to this simple house, a fragment of Sherif's past. I wandered up and down for a while, taking photos.

It seemed strange that the commemorative street had disappeared. But then I remembered the election of the new mayor, who was from Erdoğan's AKP party. Was it a simple matter of a new incumbent in the mayoral office staking his claim? Or was it part of something larger, the obliteration of reminders of those who supported Kemalism and the founding of the secular Republic? It wasn't until later that I learned this wasn't the first time Sherif had been sidelined by the winds of politics in Turkey.

Perhaps there was no single event, no one moment where Muzafer Sherif's childhood was snatched away from him. From the age of six he lived in an empire, and then a country, that was constantly at war: beginning with the Italo–Turkish War in Libya in 1911, to the Balkan Wars from 1912 to 1913, to World War I from 1914 to 1918, and then to the War of Independence, which lasted from 1919 to 1922.

As a typical child of the late Ottoman Empire, Muzafer grew up playing games, reading books, and listening to his mother's stories of trickster Nasreddin Hodja. At school he learned how to read and write Arabic and to recite the Qur'an. But slowly, those wars that had until then seemed remote, just the mutterings of his father and uncles at the tea house, would have come into sharp focus and thrust themselves into his consciousness.

Across the Ottoman Empire, rising nationalisms were dissolving loyalty to the sultan, which had previously united the melting pot of different ethnic religious and cultural groups in Anatolia. Mistrust between Ottoman Greeks and Turks surged after Greece declared war on the Empire during the Balkan Wars, and after the Empire's humiliating loss of the prized city of Salonica in 1912. Beginning in 1913, the Ottomans began the forcible expulsion of Greeks and, later, Armenians, who had lived in the area for generations and whose family history, language, memories, and loyalty were bound up in their town and neighbourhoods, but were now regarded as traitors. As part of a government push, they were driven out of their homes and businesses or rounded up, their properties looted, their churches destroyed. In the winter of 1916, when Muzafer was ten years old, the town crier, accompanied by a small boy beating a drum, made his way through the streets of Ödemiş in a flurry of snow, announcing the edict that began with an account of the subversion and treachery of the Armenians and ended with the news of their deportation. Five days later, amid terrible lamentations and wailing, 1,500 Armenian families and their possessions were loaded onto donkey and oxen carts. Under the escort of gendarmes, they were marched out of Muzafer's home town.

But almost as soon as the familiar faces of neighbours, shopkeepers, and friends disappeared, a wave of strangers arrived to take their place. Anatolia was engulfed after the Balkans War, with almost half a million refugees arriving between 1913 and 1918.

From what I saw at the Ödemiş cultural museum, any reference to local Greeks and Armenians, who had been part of the area's culture for centuries, had been edited out of the story of the town. In much the same way local Armenian churches have been converted into mosques, the museum told a story about the town that ignored uncomfortable truths.

In villages and towns across Anatolia, differences that had been submerged by generations of intermarriage, the practicalities of business relationships, and the interdependence of neighbours suddenly resurfaced. Friends and neighbours who had lived in the same area for generations now viewed one another through the prism of religion and politics.

History books describe these dramatic changes, and some memoirs by leading writers such as Halide Edib and Irfan Orga touch on them, but none can tell me specifically what villagers and townspeople in places like Bozdağ or Ödemiş thought of what was happening: if they supported the expulsions or if they felt outraged or ashamed. So we can only guess that people such as Muzafer and his family, who counted Greek and Armenians among their employees, shopkeepers, neighbours, and friends would have felt. They would probably have been torn between violent love for their country and hatred of traitors, and affection and fellow feeling for their community members. For some, patriotism won out every time. Others remained conflicted and troubled, and spent a lifetime trying to make sense of it. Many simply tried to forget.

For Ottoman children, this seismic shift in loyalties was reflected in school readers. In Muzafer's first couple of years at school, before 1913, students learned an idealised version of the Ottoman Empire, built on unity, equality, and cooperation among people from a wide variety of ethnic and religious communities. At the same time, Turks were presented as superior for having established the Ottoman government, for having the longest history defending the Empire, and for being the largest and most powerful group in its borders:

> These sacred lands, on which various ethnic groups and elements like the Turks, Arabs, Albanians, Bosnians, Kurds, Laz people, Georgians, Greeks, Armenians, Jews, Bulgarians live with a language and are unified under a common benefit are called homeland. The composition of these elements living under the same rule are called "nation" (millet). This homeland is Ottoman; our common nation is the Ottoman nation … love your homeland and nation much more than your life, and live proudly under the honorable Ottoman flag.

Instead of playing games, patriotic schoolchildren, like Muzafer and his brothers, were expected to practise military skills so as to prepare for adult military service. It was without question that they would engage in the defence of their homeland.

After 1913, Muzafer's schoolbooks told a dramatically different story. The Empire had lost more than half of its territories in the Balkan Wars. Traumatised by these losses, the Young Turks, in their nation-building, organised history lessons in primary schools that emphasised a break with the past. Schoolbooks exhorted the new generation of nationalists to mourn the loss of the lands of the Empire, to protect its territories, and to seek retaliation against those ethnic groups who had betrayed them. These texts did not censor or protect children from stories of brutality and trauma. Now many of the cultural and ethnic groups that had been encouraged to rally together under the Ottoman flag were traitors, as the passages made clear:

> They behaved like worms inside us. They joined our foreign enemies. They took ¾ of our motherland and wounded our dear mother. They killed thousands of, hundred thousands of suckling babies, raped our women, and, these monsters even raped our little girls. Turkish child! Remember these enemies who were previously worms, and who transformed into snakes now. The blood of your grandfathers is shouting: "Turkish child! Take your revenge!" Do not ever forget the words Greek, Bulgarian and Serbian and those who want to behave like them!

Increasingly, with war, the battle for new territories, and rising nationalisms in both Greek and Turkey, the classification of Ottoman subjects according to race and religion gained new power. The ties that bound the multiethnic groups were unravelling. The question of which group you belonged to — Armenian, Greek, Turkish, Jewish, Christian, Muslim — was no longer straightforward. How did you identify yourself? Whose side were you on?

In 1914, when he was nine, Sherif's school closed its doors — like so many during World War I. For his mother, Emine, keeping an eye on five children with no school to occupy them would have been a worry.

For Muzafer, no school would have meant freedom and a chance to explore. He must have been a cocky and adventurous boy. One day, he and a friend set out for Smyrna, 60 miles away; it was too far to walk so I imagine they caught an occasional ride on a donkey cart, following the railroad track for 60 miles until they reached the city.

It's easy to see why a boy from a small market town would dream of an adventure in Smyrna. It faced outward, to the sea, towards Europe on one side and Asia on the other. Behind it, the hinterland, which included Ödemiş, seemed a world away. Smyrna was known as the 'insolent' city: the town's valis, local governors, were often men who had been banished from Constantinople for their opposition to the sultan. Inhabitants of the city were skilled at circumventing the sultan's strictures and enjoyed religious and economic freedoms not at all typical of the rest of the Empire. Even the local newspaper was careful in its reporting of the city's social activities to avoid angering the sultan. Its tolerant atmosphere and prosperity was a contrast to other, war-ravaged parts of the country. On Smyrna's streets, traditional Ottoman subjects wore turbans, fashionable Turks wore the fez,

Greeks wore flat-brimmed caps, and Americans favoured broad-brimmed straw hats. Baggy Zeybek–style pants were banned: at the railway station, men queued to hire European–style trousers for their visit to the city. People who considered themselves sophisticated eschewed Turkish and spoke Greek and French.

Above the cries of sherbet and fruit sellers shouting their wares, Greeks yelling over games of cards through the open doorways of cafés, the ever-present barking of dogs, and the sputter and roar of motorcars, the bells of St Photinia, taller than the minarets of the mosques, rang out across the city at the same time as the call to prayer.

On Frank Street, bakery windows were full of French cakes. Muslim women wore transparent veils or no veils at all, and close-fitting clothes that showed their figures. At Bon Marché and Petit Louvre department stores, young women entered iron-grilled elevators that took them upstairs to be fitted in the latest fashions copied from French and American magazines. But although the city's tolerance of diverse religions, cultures, and nationalities might have been part of its charm and was the secret of its wealth, it was also part of its downfall.

The Kordon, the long street that ran along the quay, with its marble-fronted houses facing the water, was crowded with camel trains loaded with cargo. Some caravans travelled with as many as 1,500 camels, bringing in Eastern luxuries — carpets and rugs, silk and cotton, baskets full of lemons — and taking back Western goods — sacks of spices, watches and clocks, Singer sewing machines. The bay bristled with ships and sailing boats, steamboats and caïques. As night fell, the traders packed up and the quay became a promenade. The pale marble façades of the buildings glowed against the night sky. The cafés below lit up, and music floated out across the water as far as whichever battleship was moored out there — a fusion of Greek and Turkish music they called rebetiko. Wharf workers, tough men who hauled sacks and carried heavy loads all day, played cards with sailors who had come ashore for the night. Above them, on the café walls, hung portraits of the Greek king and queen.

This mixing and intermingling — of languages, religion, and culture — was likely heady stuff for a curious mind. To nine-year-old Muzafer, Smyrna was still a magical place.

But his happy wandering through the streets with his friend that day was cut short. He was grabbed from behind by his uncle, a doctor in Smyrna, who shook him furiously. What was he doing? Didn't he know how many people were out looking for him?

This time, Sherif returned to his family unscathed, his head full of the magic fairytale city, but he wouldn't always be so lucky.

# 11
# Burning Memory

Once I boarded the train at Ödemiş, I couldn't wait to get to Izmir, the city that had been known as Smyrna when Muzafer Sherif was a boy. Ödemiş had felt like a secret that I couldn't crack, but I had read a lot about the almost legendary city of Smyrna during the period that Sherif had lived there, and I was excited to see it. In addition, one of the academic experts in Turkish psychology I had emailed was based at a university there, and I was hoping to meet them and hear their interpretations of Sherif's research against the backdrop of Turkish history.

The train sped through idyllic scenery: green fields, with the sun burning off the haze of clouds, and women bent over crops, their bright headscarves fluttering, a rainbow of irrigation water arcing in the sky behind them.

It passed through Şirinyer, where Muzafer Sherif had been a boarder at the International College of Smyrna, founded by Christian missionaries. Up and over the squat NATO building that stands on the busy corner of the street in Izmir, over the tops of the trees, you can see the clock tower of MacLachlan

Hall. In 1918, thirteen-year-old Muzafer Sherif arrived with his trunk, excited to pass through the ornate gates, up the driveway lined with cypress, and into one the most exclusive schools of the Ottoman Empire. Boarders came from as far away as Greece and the Aegean Islands; day boys arrived by steam train each morning from Smyrna, getting off at the place known to locals as Şirinyer but the missionaries had called Paradise. The school was 250 feet above Smyrna, and a mile and a half away, in a lush green valley surrounded by hills covered in wildflowers, with snow-capped mountains to the east. Situated on choice farmland, the property included a vineyard, orchards, a vegetable garden, and a large field for crops. Each of the more than a dozen faculty members had their own house that accommodated their families and servants. Beyond the school wall, in a dip in the valley, ancient Roman aqueducts spanned the river and, nearby, a railway station connected Paradise to Smyrna. It was a place where a boy might forget about war.

Today the farmland is gone, the hills are covered in houses, and Şirinyer is a suburb of Izmir. The school is fenced off, swallowed up by NATO, who use it as part of their headquarters. As the train passed, I scanned the skyline, the rooftops over the spires of minarets. I'd seen a photo of the school's distinctive clock tower rising above the low red roofs of surrounding barracks, the dark green of the pencil pine rising up like the tip of a paintbrush beside it. But the train was going too fast; Şirinyer and the clock tower flashed past.

School photos of the period showed young men in the school's uniform of suits with crisp white shirts and ties, seated around a long table set with silverware and crockery in a formal dining room, being waited on by servants. All the students came from affluent families, and for some Muslim parents such as Muzafer Sherif's, a wealthy 'foreign' school with money to spend offered a better education than a local Ottoman one, even if it was Christian.

Many of the students who attended school with Muzafer would go on to prestigious universities in Europe and abroad, including Switzerland, England, and America. Others wouldn't live that long.

I knew of course that Izmir was a transit point for refugees on their way to Europe, but before I arrived, I wasn't prepared for the scale of it. Turkey's third-largest city has always been known as a progressive place with a relaxed atmosphere. But it was crowded: the streets were jammed with traffic — drivers leaned angrily on their horns — and the footpaths thronged with ceaselessly moving crowds. In the hotel courtyard, turtles swam in circles in a tiled blue pool as if trying to find an escape.

My room wasn't ready, so I headed for the Kordon, the seaside walkway fringed with palm trees that had been such a magnet back in Sherif's day. Izmir sits on the sweep of a bay surrounded by mountains. Over centuries, its houses and buildings climbed from the flat land of the shore, up the hill, towards the castle at the top. But the geography of Izmir confused me. In my mind, all streets led to the waterfront, yet there was a busy road running parallel to the sea, separating it from the rest of the town.

Instead of the cobbled walkway along the waterfront, the Kordon was now a concrete path lined with lawns of trimmed grass, and beyond that, a jumble of pastel apartment buildings and hotels facing the water. The promenade was largely deserted. Turkish flags jangled against flagpoles in a relentless wind that flattened my hair and made my eyes water so that I turned my face away from the blue of the bay. Then I noticed the small family groups: mainly women sitting on the grass with surprisingly quiet small children. There were more sheltered spots to sit, but they sat in the full blast of the wind, facing the water and the sight of Lesbos, just 6 miles away. Some talked, but mostly they were silent — looking hopeful or hopeless, I couldn't tell. It was a tableau repeated the length of the Kordon. As I passed, I glanced over at each group trying to guess where they had come from — Syria? Afghanistan? Iraq? — thinking on the circumstances that had landed them here.

I watched the next stage of their journey on CNN in my hotel room: people crowded into rubber dinghies, the men pushing boats from the shore, the parents holding their children close, the boats veering in circles before straightening up and hopefully heading straight towards Lesbos. Down the coast, just a short distance from Izmir, bodies continued to wash up on the shore.

Interspersed with news of the refugee crisis were updates on the upcoming Turkish election and President Erdoğan continuing crackdown on journalists and intellectuals critical of the government. Of course few of the Turkish academics I had emailed had answered. I knew from the news headlines what was happening in Turkey before I travelled, but suddenly this project, wandering the streets of Izmir looking for echoes of the past, seemed shallow and frivolous in the face of what was happening in the present.

The mistrust between the Greeks and Turks that grew after Greek independence and Greece's fight against the Ottoman Empire in the Balkan Wars reached new heights after Greece entered World War I on the side of the Allies in 1916. At the same time, Committee of Union and Progress leaders publicly blamed the supposedly traitorous Armenians for the disastrous Caucasus campaign. Underneath its cosmopolitan surface, the fault lines of race, nationality, and religion in Smyrna were widening.

At Sherif's school, the student population had ebbed and flowed in concert with the tide of politics and war around it. By 1918, most of the Christian Armenian boys, who had made up the majority of pupils since the school was founded, had disappeared. Scapegoated for the Caucasus defeat, 600,000 to 1,000,000 Armenians were marched towards the Syrian desert, where most of them perished.

During World War I, the British promised Greece territorial gains in Turkey in exchange for entering the war on their side, and after the war they made good on the deal. After the armistice was signed in November 1918, Allied troops sailed into Constantinople and occupied the city, and Greece was granted the western regions of Anatolia, including Smyrna. An Englishwoman in Smyrna called Emily Holton kept a diary and wrote about the arrival of the first British ship in port after the armistice:

> The people were frantically excited and caught the captain up when he stepped on shore and cheered the rest of the officers. The town was bedecked with flags mostly Greek & they were waved around by individuals … All the villagers sported little Greek flags and walked about the streets all

> night singing, cheering & shouting 'Liberty'. The church bells rang at intervals and there was service in the churches and a procession with the pictures of the saints at nearly midnight. ... Greek flags hang out of private houses. It is a wonder the Turks are so quiet with all this triumph flaunted in their faces.

At Muzafer's school, the headmaster tried to keep politics outside the gate, but the faculty was divided along their support for the Greeks or the Turks, and feuds simmered between teachers. Among the students, former friends looked at one another through new eyes. Selma Ekrem described in her autobiography the tensions between Greek and Turkish girls at her college in Constantinople at news of the landing of Greek troops in Smyrna:

> At college mournful groups of Turkish girls pored over the newspapers ... The occupation of Symrna by the Greeks had revolted us ... Turkey would cease to exist — a greater disaster than the Great War had fallen upon us ... We could not bear to look at the other girls, those who were not Turks and who lived their happiest days in our blackest ones.

As a fourteen-year-old boarding-school student, Muzafer Sherif was there to witness the occupation of Smyrna. On 15 May 1919, he stood among the crowd, who had gathered to watch the arrival of Greek troops, on the quay in Smyrna. An observer described the quay as a sea of red fezzes, the headgear that identified the wearer as an Ottoman subject. Some of the crowd were sombre; others were celebrating. For Ottoman Muslims, the arrival of an invading army was a cause for bitter humiliation and anger. The Greeks in Smyrna were ecstatic at the troops' arrival, waving pictures of Greek prime minister Eleftherios Venizelos and showering the soldiers with flowers. Church bells pealed in celebration and bands played the Greek national anthem. To the Greeks, it was an exhilarating first step towards recovery of the Byzantine heritage; to the Ottoman Muslims, it was clear the rejoicing Greek crowd would soon join their soldiers to fight the Turks. With Smyrna as the soldiers' base, Sherif and his fellow students would be confronted daily with the sight of the white-skirted foreign army, the everyday reality of the occupation of their country, through interactions with the enemy and those who supported them. For many on the quay that day, the ecstatic reception by the Greek crowds stoked the fires of anger and nationalism.

The festivities didn't last long. The troops began rounding up Turkish soldiers and citizens, marching them up and down the quay, forcing them to shout 'Zito Venizelos' ('Long live Venizelos'). Sherif's daughter Sue outlined Sherif's experiences for me. 'My father heard shots being fired, and he and his friends rushed to see what was happening. They were boys, it seemed like an adventure — a bit like your Aussie boys, who joined the army to see the world and when they got to Gallipoli, it was like, "Oh my God!" My father said it was horrible. No wonder he didn't talk about it.'

Those who refused to cheer for the Greek prime minister were tortured and bayonetted in front of a large crowd of cheering locals, many of whom joined in. Caught up in the violence, Sherif stood frozen as soldiers murdered people all around him. He later told a journalist it was a 'miracle' that he wasn't killed.

According to reports, a few hundred died on the Smyrna quay that day, and many were taken away and tortured. Almost immediately, the reprisals began, with violence, murder, and looting. The arrival of the Greek troops in May 1919 splintered the Empire and galvanised the Turkish resistance effort, the Turkish National Movement.

By June, the Greeks had pushed inland towards Sherif's home town of Ödemiş.

On a hillside outside of present-day Izmir, a huge bust of Mustafa Kemal Atatürk, seemingly carved from a mountain, Mt Rushmore–style, looms out of the rock, casting a shadow across the landscape. Atatürk's spirit haunts the city. It was in this area that the first armed clash of the war took place. Six months before the Greeks landed in Smyrna, the Allies occupied Constantinople and met with no resistance. But in Smyrna, it was a different story. In a village near Ödemiş, a group of armed villagers organised themselves to fight the invaders. When I read this, I remembered the statue of the man with the rifle I'd seen in the street at Bozdağ, and I looked him up.

Poslu Mestan Efe was a brigand and a local hero who commanded an unofficial group of militia that attacked invading Greek forces in 1919. His actions allowed Mustafa Kemal time to secretly knit together resistance groups across Anatolia and organise and establish a regular army. These were the wild days — the Empire was crumbling but nothing new had yet taken its place. The brigand distributed arms to local villagers and, advancing on horseback from the foothills of the mountains, they ambushed

Greek troops in a series of attacks in and around Bozdağ. The spark of the Resistance movement that would eventually oust the Greeks and threaten the occupiers of Constantinople was lit in the village where Muzafer Sherif was born.

It goes some way to explaining why, within a year of the Greek invasion, as a fifteen-year-old schoolboy Muzafer Sherif had joined the Young Turk movement and the nationalist cause.

When night came, the hotel corridor whispered: bags rustled, voices murmured, cell phones pinged. Outside, the streets were choked with people. Men and women clutched black plastic bags filled with belongings, held the hands of children old enough to walk and carried those too young. They swarmed across the busy roads, dodging traffic, heading purposefully towards their destination. The sight of them filled me with anxiety and dread. They were the ones who had received the call they had been hoping for. A boat was waiting. First stop, Greece. If they made it. Beyond that, who knew.

In between serving food, waiters in cafés engaged in protracted negotiations with patrons at the tables and then went to stand behind the counter, talking urgently into their phones. Barber shops, tobacconists, and newsstands did a brisk trade in fluorescent vests. I stopped at a former menswear shop, where mannequins of a man, a woman wearing a hijab, and a child in orange lifejackets were posed in the window. I pointed at one of the vests and asked the shopkeeper, 'How much?' but he waved his arms. 'No journalists!' he cried angrily.

On makeshift tables in the side streets around the neighbourhood of Basmane, children sold homemade waterproof pouches, each made from a sandwich bag, a balloon, and a piece of string, to keep passports and identity papers dry and buoyant in the water. I wanted to leave at the same time that I felt guilty I could.

A former carpet merchant's house on the quay at Izmir had been converted into the Izmir Atatürk Museum. The museum's rooms and displays told the official history of the birth of the Republic of Turkey. In one room stood eerily lifelike mannequins of Atatürk and his commanders in military uniform, poring over a map, presumably plotting their next move and discussing the progress of their war.

In another room, I was excited to see a photo in the museum of Mustafa Kemal Atatürk at a boys' high school, surrounded by a large group of students. They were positioned in a triangle, with Atatürk at the head and the group fanning out on both sides of him. The foreground was crowded with boys who looked like they had shouldered their way in front of the camera to get a photo of themselves with their leader. I leaned in and peered at each boy's face, but I didn't recognise Muzafer Sherif among them. Kemal was Sherif's hero, as he was for most Turkish boys of his generation. They wore a picture of him on an armband and followed the progress of nationalist forces with great excitement, despite the strictures of their school's headmaster.

The Turkish nationalists finally repulsed the Greek army in 1922. Their revenge, when it came, was terrible.

By 9 September 1922, nationalist forces had successfully pushed the Greek troops back, and the last of those who had occupied Smyrna since 1919 withdrew. Those that were captured were forced to parade through the streets shouting 'Long live

Mustafa Kemal!' amid jeering onlookers. Refugees fleeing both armies engulfed the city: Christian, Muslim, and Jewish families carried what belongings they had been able to salvage from the razing of their villages. Rape, looting, atrocities, and savagery were reported on both sides.

The city of Smyrna was in chaos. The streets were choked with refugees; the roads into town were littered with the bodies of dead and dying and the carcasses of animals. Smoke from burning villages smudged the skyline.

Meanwhile Mustafa Kemal and his cavalry escort drove into town on 10 September in an open car draped with olive branches. Mobs took over the streets. The Greek archbishop, who had welcomed the Greek troops to the city three years earlier, was lynched by a Muslim crowd.

As fleeing Greek troops fought last-ditch battles with Turkish soldiers nearby, Sherif's headmaster heard 'disturbing rumors' of thousands of mounted Turkish brigands gathering near Ödemiş on their way to Smyrna. Some stopped just out of Smyrna, at Muzafer's school in Şirinyer, on Monday 11 September. Despite the American flag that flew from the tower of the hall signalling the school's neutrality, the brigands began looting one of the school buildings. In his memoirs, Sherif's headmaster Alexander MacLachlan wrote about how when he went to stop them, he was ambushed. As a crowd of anxious boys and staff watched with horror from an upper-storey window — was sixteen-year-old Muzafer one of them? — the men toyed with him, stripping, beating, and stabbing him. One student ran out, pleading for the headmaster's life, but they ignored him, and were only stopped when a passing Turkish officer intervened.

Fearful of the retribution of the nationalist army, the tens of thousands of Greeks and Armenians who poured into Smyrna gathered on the waterfront, where they were likely reassured by the Allied warships at anchor in the harbour. Meanwhile, many of Smyrna's foreigners — British, French, Italians, and Americans — were safely boarded onto twenty-one Allied battleships with orders to protect their own citizens.

By 13 September, different Christian parts of the city were on fire. The Armenian quarter was the first to go up. Other Christian districts quickly followed, and the fire merged into a wall of flame three kilometres long and thirty metres high. As the fire roared through the streets, buildings collapsed, and horses and camels screamed in terror. The shops and department stores, the churches, theatres, hotels, factories, coffee houses, and consulates were destroyed. Thousands of people fled to the waterfront to escape the advancing fire, joining the huge crowd that had been gathering there for days. But now they were trapped between the city and the water. When night fell, they were attacked and robbed by brigands. Some escaped by boat; many perished trying, murdered on the quayside or drowned in their attempts to reach the safety of Allied ships in the harbour. The smell of rotting carcasses and bodies that bobbed on the water was so overpowering that Mustafa Kemal — later to be known as Atatürk, or 'Father of the Turks' — moved from the waterside mansion where he had been staying to one downwind of the bay. The frantic screams of the people on the quay could be heard for miles. The wall of fire was so tall that it was said that it could be seen by the monks on Mount Athos, on the other side of the Aegean Sea.

Finally, on 16 September, Kemal allowed Allied ships to evacuate survivors, except for thousands of Greek men aged eighteen to forty-five, who were separated from their families and deported as forced labour to rebuild villages in the interior after the war. Over 200,000 people were ferried to Greece. One of the conditions of the peace settlement signed in Lausanne the following year was the forced exchange of remaining Greek Orthodox residents of Anatolia for Muslims living in Greece. Turkish-speaking people of Greek heritage whose affiliation and family history were rooted in Anatolia rather than Greece, and Greek-speaking people of Turkish heritage with strong links with Macedonia, were swept up in the exchange and forced to resettle in places where they were most often regarded as unwelcome strangers.

Sherif, who was in his final year of schooling in 1922, was likely safe from the worst of the atrocities, but he would have seen many of them first-hand. Smyrna was in ashes. Anatolia had been devastated by the War of Independence. The expectation was that Sherif and his generation would rebuild it.

The fire destroyed all reminders of Smyrna's Ottoman cosmopolitan past. The mansions and clubs, hotels and cafés, were gone. The Levantine district, the Greek and Armenian quarters, were wiped away. In its place was a new city, now given its Turkish name: Izmir.

The fire that obliterated the multiethnic port city of the Ottoman Empire has been swallowed up in silence. In his victory speech to the Turkish Grand National Assembly just weeks later, Mustafa Kemal recounted how Turkish nationalists won the war but made no mention of the burning of Smyrna. Five years later, in his famous six-day speech in 1927 about the founding of the Republic, there was also no mention of the fire. Mustafa Kemal blamed the loss of the Ottoman Empire on the nationalisms and treachery of ethnic and religious minorities. For Kemal, a strong and unified country was one with a homogenous population and a new, shared history that helped people feel a sense of belonging. And that meant telling a new story of the vanquishing of a common enemy that stressed the liberation rather than the destruction of what was then Turkey's second-largest city, and the triumph of Turkish nationalists over foreign powers.

Sociologist Biray Kolluoğlu-Kırlı points out that the Great Chicago Fire of 1871 was equally catastrophic and produced an enormous number of literary, scholarly, and popular accounts of the event and its aftermath. Yet by 2005, when her paper was published, the Smyrna fire was still swallowed in silence, with 'not a single Turkish novel, film or memoir' referring to it. Any accounts of the Smyrna fire would mean moving into the dangerous territory of voicing who was responsible for setting it, and who was responsible for letting it burn.

At the museum I moved from photo to photo, from one glass case to another, peering into the rooms where Atatürk slept, read books, took baths, and had his daily shave with his personal barber. But nowhere in the building was there a reference to the event that caused its original owner, the carpet merchant, to 'surrender' the premises. The catastrophic fire that destroyed the city of Smyrna was represented by a single blurred photo without a caption that was part of a short film about the Republic, playing on a loop. The picture was taken from the sea, where black clouds of smoke billowed upwards from the city. The next frame showed

Turkish soldiers standing in the ruined and still smoking remains of buildings as if surveying the damage. I kept looking to see how the fire would be explained, and importantly who would be blamed for it, but there was nothing of the sort. The gaps in the museum's exhibit mirrored those in official histories of Turkey.

It's one thing for official histories to be edited, but what does that silence do to people who were there? What do they do with the traumatic memories, emotions, and experiences, especially ones that might contradict official accounts?

Muzafer Sherif never talked about his experiences of this violent and traumatic ethnic conflict in detail with his family. Perhaps that's not surprising — perhaps, like so many others, he didn't want to remember. But he did write about it once, when he was sixty-one. In a 1967 book of essays about his work, he foregrounded his childhood as the inspiration for his intellectual development:

> I will say something about this personal background. As an adolescent with a great deal of curiosity about things, I saw the effects of war: families who lost their men and dislocations of human beings. I saw hunger. I saw people killed on my side of national affiliation; I saw people killed on the other side. In fact, it was a miracle that I was not killed along with hundreds of other civilians who happened to be near one of the invasion points the day Izmir (Smyrna) was occupied by an army …
>
> It was the period of the final dissolution of the Ottoman Empire and the rise of nationalisms within the disintegrating

> empire. The rise of Turkish nationalism, which fascinated me as it did all other youngsters of my generation, resulted in the new Turkish Republic, born against great odds, against obstacles created by colonial powers …
>
> I was profoundly affected as a young boy when I witnessed the serious business of transaction between human groups. It influenced me deeply to see each group with a selfless degree of comradeship within its bounds and a correspondingly intense degree of animosity, destructiveness and vindictiveness toward the detested outgroup — their behavior characterized by compassion and prejudice, heights of self-sacrifice, and bestial destructiveness. At that early age I decided to devote my life to understanding and studying the causes of these things.

How strange that he finally broke his silence about his past in a textbook. I have been struck each time I read this passage by what it doesn't say as well as by its abstract language. Sherif positions his boyhood self as an observer and a witness, rather than as a participant. There's no hint of exactly which groups of compassionate-one-minute, brutal-the-next people he was referring to, let alone to which of them he belonged. By using abstract language and avoiding detail, he has reduced the turmoil and trauma of war to an abstract and intellectual problem — one he would 'devote his life' to studying.

Given he was revealing this information in an academic book, the fact that he didn't dwell on the personal and merely related the facts of his experience in a clinical, impartial manner is hardly surprising. It was the norm for social scientists writing in that context at the time. But it made me wonder why he mentioned his background at all, particularly at this time in his career. Perhaps he thought connecting the research more explicitly to his life would be more appealing to readers? Yet it seemed odd to me that someone who prided himself on his objective science, and especially someone so reluctant to discuss the past with his family and friends, would include it in a book. And even though Sherif frames it as part of the story of his 'intellectual development', what he describes is in fact profound personal history. Why was it that he wanted these formative personal experiences on the public record, that he wanted it known that he had a reason for his research that went beyond the professional? I had the feeling that he was both revealing and hiding himself in this account of his past.

On the walls of the Izmir Atatürk Museum, photographs attest to the transformation of what were presented as the primitive and backward ways of the Ottoman Empire to the modern and westernised Republic of Turkey. Workers toiling in fields or riding donkeys dissolved into images of machines pumping in factories and groups of young women doing calisthenics and jumping jacks.

In one glass case, Atatürk's elegant kid gloves, golfing cap, and highly polished spats are on display, as a celebration of his sophistication. But the display is more than a tribute to his taste; his European clothing was a symbol and a powerful message to the citizens of the new Republic. There is a photo of him parading down a village street in a pale linen suit wearing a panama hat, swinging a cane, while on the other side of the street, going in the opposite direction, an old man dressed in a traditional long robe, wearing a turban and full white beard, gapes as Mustafa

Kemal, as he was then known, strides by. I guessed it was taken during his 1925 trip to Kastamonu, in northern Turkey, as part of his campaign against the fez. It was famed as a conservative town, where the idea of adopting Western headgear was akin to a betrayal of religion. In his book about the outlawing of the fez, author Jeremy Seal wrote how during Kemal's brief stay in the small town, the national hero kept returning to his rooms to change outfits, and used the streets as a 'catwalk where he previewed a radical show called the twentieth century'.

With the breaking up of the Ottoman Empire, Mustafa Kemal's government began homogenising the previously multiethnic, multireligious, and otherwise diverse groups of people who had comprised Ottoman subjects. Citizens of Turkey needed a new identity, something that would cement the disparate groups now living in the region — many of them people with Turkish origins repatriated from other parts of the Empire such as Greece — into a whole. They had to learn what it meant to be Turkish.

Intent on banishing the Ottoman past and what the new government saw as its obsolete and reactionary values, Kemal embarked on a series of radical social reforms to forge a new national identity, making citizenship of the Republic the bond that would unite citizens. The people would embrace the term 'Turk', with its pejorative connotations from Imperial times, and use it as a term of pride. In the secular Turkish Republic, people would dress the same; read, write, and speak the same language; and behave in ways that distanced them from what the revolutionaries saw as the backward and superstitious ways of the Ottoman Empire. But for many, whose lives had been governed by tradition and God, the new Republic identity involved difficult moral choices. For some citizens, for example, abandoning the fez in favour of western headgear was a form of apostasy. But resistance or criticism of the new reforms were dealt with by force and sometimes violence. Those who continued to wear the fez were arrested and jailed, and some were hung.

It was the job of young nationalist intellectuals such as Muzafer Sherif and others like him to help bring about acceptance of the changes that came with the new Republic using not force or violence, but the latest science. And for that, he would have to travel to America.

On every street corner, at busy intersections or on quiet streets, Atatürk's face stared down from banners and flags. The wind flapped the flags bearing his face in a strange kind of game of now-you-see-him-now-you-don't. Sometimes he was depicted wearing a black wool hat, wider at the top than at the bottom, and a military uniform; in others, he looked suave in tails and a wing collar — Ottoman officer and military man one minute, suave westerner the next. Perhaps it was my growing discomfort at being here in a kind of disguise myself, but I felt as if those blue eyes were watching me wherever I went.

I had come to Turkey looking for the personal origins — if that was not too strong a term — of Sherif's tribal war and peace research, and I was attuned to these echoes of history. The more I read about the events that led to the founding of the Republic of Turkey, and particularly those around Ödemiş and Smyrna, the more I could see the notions of friendship and betrayal, the dissolving of group loyalties, and the forming of new alliances and identities as central themes in the country's history at this time. But how did this history intersect with Sherif? How was his own sense of selfhood altered and reformed during this time? Did members of his family share his nationalist zeal and embrace the change, or did they resist it? And how did he feel about it?

I realised, as I was preparing to leave Izmir, where the city of Smyrna seemed little more than a place of my imagining, that I would never really know how events might have shaped Sherif — all I had were documents and glancing accounts, inferences and hints. Perhaps the idea of a thread from past to present is fanciful, as if you can ever really plumb history, or the people who dwelt in it. Who knows. Even if I had more to go on, an autobiography perhaps or a cache of letters, who is to say he was any more of a reliable narrator of the past than I have been?

But now I was too far into the story of Sherif and the lost boys to abandon it. I was heading to America and as I watched Turkey dwindling below me out of the plane window I began to think that America may provide more of a clue — or at least, a conclusion to the narrative I was pursuing.

# 12
# America and Back

When he arrived at Harvard in 1929, Muzafer Sherif knew little about the university, except that one of his heroes, psychologist William James, had studied there. Harvard was exclusive, wealthy, upper class and, having been at the forefront of the eugenics movement, it had until relatively recently been teaching its students the science of racial superiority.

Sherif's expectations of America would have been shaped by the films he saw in Istanbul's movie houses — glamorous women, sophisticated nightclubs, limousines, jazz. But a month after he started classes in September 1929, the stock market crashed and he articulated into a very different world.

Muzafer Sherif's Harvard teachers and classmates noted that he had strong opinions and thought nothing of sharing them. He probably inherited this trait from his father, who was known for his single-mindedness. Serif Effendi started life as an illiterate farm boy, but taught himself to read and write as an adult and transformed himself into a landholder and then a successful businessman. He was an intelligent man used to getting his way.

Sue Sherif told me that while he ruled over his family with an iron fist, he also provided all his children, including his daughter, with a university education and supported them financially in their studies overseas, even after it became clear that none of them were going to obey his wishes to become medical doctors, a highly prestigious occupation in Turkey at the time.

The fact of his children's university education in what was an agricultural community was extraordinary, and even today makes the family a local legend. Muzafer Sherif's daughters found that at their grandparents' gravesite at Birgi, a small town near Ödemiş, coloured strips of cloth fluttered on the fence, tied there by pregnant women so their prayers for children as well educated as Serif Effendi's would be answered. With the weight of his family's and his country's expectations on him, it wasn't surprising that Muzafer Sherif demanded a great deal from himself.

But I wondered if this perception of Sherif at Harvard as blunt and outspoken could also have been a clash of cultures. Listening to recordings of him, Sherif's heavy accent and emphatic way of speaking, even when it was about something light-hearted, often made him sound more forceful than perhaps he was.

Sherif arrived in the United States unnoticed by immigration authorities. According to his daughter Sue, he boarded a freighter in Egypt that travelled via the Canary Islands and through the Caribbean. For reasons the freighter's captain didn't explain, he bypassed officialdom and sailed past New York, stopping instead at Providence, Rhode Island, where all the passengers disembarked. It was 9 August 1929, and he had just turned twenty-four.

He might not have had much money, and his arrival in America was unobtrusive, but Muzafer Sherif made up for it with his conviction that he was destined for great things. He was a privileged, well-educated young man, already marked out as a future candidate for his country's intellectual elite. In his years at college in Smyrna and then at university in Istanbul, his teachers saw great promise in him, wrote him glowing recommendations to American diplomats in Turkey, and used their influence and connections to get him a Harvard scholarship.

But after he left for Harvard, the letters I've read that crisscrossed the ocean between his supporters in Turkey and his mentors in America hinted at problems of 'temperament', as if he needed careful handling. From what I could gather, in America Sherif experienced intense emotional highs and lows, alternating between soaring ambition and crushing self-doubt.

On the face of it, when he arrived at Harvard in 1929, Sherif fitted in. Photos of him taken in Cambridge at the time show him looking handsome in his suit, tie, and crisp white shirt, his black hair brushed back from his high forehead. He was self-assured, and quickly made friends with fellow student Hadley Cantril, who had powerful connections: at Dartmouth, Cantril had shared a room with Nelson Rockefeller. Over 6 feet tall, Cantril — who had abandoned his first name, Albert, for his more distinguished-sounding middle name — was, with his intense blue eyes, good-looking, charming, and 'pathologically ambitious'.

But despite looking and acting the part of a Harvard student, Sherif felt as though he didn't belong in the rather snobbish and exclusive club that was the university. He sensed that some students and faculty looked down on him and felt he had to work hard to be taken seriously by 'blue bloods', whom he said had a hard time believing that 'a Turk could read or write'.

Certainly public opinion in America at the time was anti-Turk. The animosity had been fed by reports from missionaries in the American press of the forced marches into the desert and slaughter of Armenians. While the attitudes of American diplomats and government officials towards Turkey shifted dramatically in light of Mustafa Kemal's zeal to 'civilise' the country, the press and the American public took decades longer to change. In the American imagination, Turks were often brutal savages who enjoyed killing, rape, and torture. A few years before Sherif arrived at Harvard, Turkish journalist Ahmet Emin Yalman wrote about how he and a friend travelled to Maine on holiday from New York and locals, who heard that Turks were coming, installed new locks on their doors and reinforced security at the local jail.

One of Sherif's mentors — Beryl Parker, an American academic in Turkey — wrote that whether the racism Sherif said he experienced was real or imagined, it had a powerful effect on him. Sherif arrived at Harvard expecting to take what he saw as his rightful place in a pantheon of the great in the field of psychology, but he was also sensitive about what he saw as his shortcomings — including his lack of background in maths and science. Yet instead of feeling embraced and welcomed, and encouraged to fill any gaps in his knowledge, he felt he had to fight to be taken seriously. Luckily, Sherif's key mentor at Harvard, Gordon Allport, had taught in Constantinople, and he sensed that under Sherif's arrogant exterior he was anxious about his ability to perform. In a letter to Sherif's former university teacher in Istanbul, Allport wrote, 'He is … inclined to work too hard, and to be too serious'. He had an 'emotional temperament that led him occasionally to feel discouraged. He feels that he is not doing enough to justify your confidence in him. In my judgement, however, he is diligent and conscientious, making the most of his opportunities, and should in time prove that your confidence and favors were well placed.'

But Sherif's habit of following his interest and reading across a wide range of disciplines also brought him in conflict with some teachers, who felt that he should be focusing more exclusively on the discipline of psychological science. Sherif's interest was already in a wide-ranging sociological brand of psychology that could be used in the service of nation-building and educational reform. What he got was an irritatingly narrow physiology- and laboratory-based science course that often involved experimenting on rats. Dismayed by the form of social psychology he found at Harvard, with its 'fragmentary and piecemeal state' and 'lack of perspective', he began casting around for an alternative framework for understanding human social behaviour. When he came across the work of German-Jewish Gestalt psychologist Kurt Lewin, with his interest in an individual's interaction with their culture and context, it was as though a window had been opened on a stuffy room: 'Lewin's work appeared at that time like a fresh breeze,' he wrote.

As soon as he had finished his master's degree in 1932, Sherif set out to return to Turkey, but on the boat he made a sudden decision to commit himself to studying Gestalt psychology and disembarked to travel to Berlin. There he could learn the German that would unlock so much of the psychological literature, as well as attend classes by Wolfgang Köhler, one of the founders of the Gestalt movement at the University of Berlin.

He arrived in Germany broke. At Harvard, he'd had to borrow

living expenses because he had spent his annual allowance — over $400 — on books, and he left debts behind. Still, he wrote to Allport, asking him for a loan to fund six months of study in Germany. 'I think you want to help me wherever I may be,' Sherif wrote. 'And I hope your interest in me will continue forever.' After Germany, he would return to Turkey, work on a PhD thesis, and then return to Harvard in a year or two and submit it, he told Allport. He would develop a new kind of social psychology: 'Sometimes it seems to me that it will be at least as good as any existing system in social psychology. Would you say that was an assertion of a person with stupid megalomania?' Then, with Allport's help, he planned to publish this work: 'It is a nice dream in me to have a Turkish name in social psychology. Please do not think that this is the only value for me. Social psychology deeply fascinates me because I intensely love and hate man as I intensely love and hate myself.' Exactly what he loved and hated about himself and his fellow man, he didn't say.

Germany had long been the training ground for Turkish academics, teachers, and military officers, and soon after he arrived he found a community of Turkish students in Berlin. He boarded with a 'Hitlerist' (before it was known what the extent of that term would come to mean) family, practising his German around the dinner table in the evenings, and attending classes by leading Gestaltists at the university. With no money, he wandered the streets looking for free events and gatherings, and was fascinated by the political rallies with massive crowds roaring their approval of their new chancellor, Adolf Hitler: 'Ten days ago I went to see Hitler at the Lustgarten. There were around 100,000 men and women, a real mass meeting. Believe me, sir, my stay here is worth all the suffering I am enduring,' he wrote to Allport. He must have wondered, as he sat at the dinner table in Berlin and his landlady served him an extra helping, how the same kindly woman could stand in a crowded square and shout her support for the hateful words of Adolf Hitler.

It's a sign of Sherif's charm and Allport's belief in him that Allport did continue to support him, sending him money and working hard behind the scenes to raise funds over the next eighteen months for Sherif's return.

In Turkey in 1930, a protest against the government had made national news when an armed crowd protested against the secularisation of the state and called for the restoration of religious law. A policeman fired on the crowd, triggering a riot, and the crowd turned on the police and beheaded the commanding officer. It wasn't the first rebellion against the government, but it was the most shocking, because it took place not in a remote region, where people had limited access to education and religion had a much firmer grip, but in Menemen, close to the city of Izmir and the heartland of Kemalism. It was a watershed moment. The government realised it had to change tactics — instead of imposing change by force, it needed new methods for convincing and educating people on the value of accepting Kemalist ideology, and one of the most powerful tools was the education system. When he returned to Turkey, Sherif was employed at a teacher training college where he taught new teachers educational methods drawing on proven techniques in social psychology for fostering and shaping desirable attitudes that would inculcate young citizens and persuade older ones of the value and benefits of Republican ideals. In the year he worked there, Sherif developed an experiment about persuading individuals to change their perceptions.

As soon as he was back at Harvard in 1933, Sherif excitedly presented it to Allport. Early astronomers had noticed how if you looked at a single star against a very dark night sky, it appeared to move. They called this the autokinetic effect. Sherif used it in an experiment to show how you could change people's perception of facts. First, he seated a man in a completely dark room and asked him to estimate how far the single point of light moved in the darkness. Even though the light wasn't actually moving, each man thought that it was and estimated how far. But when one man was joined by a group of others and they gave their estimates out loud, the man's judgement converged with the group's. If he estimated the movement as, say, 6 inches on his own, he began to estimate it as more like 4 inches, increasing or decreasing his estimates in line with the rest. Without being aware of it, people changed their way of seeing things to more closely resemble the views of others in their group. The influence of the group was a powerful force in changing people's attitudes.

But people weren't aware of this process, Sherif argued to Allport. Our judgement, our perception of what's true and what's not, what to hold onto and what to reject, is shaped by people we identify with, our tribe. This explains how people absorbed positive and negative beliefs and racist attitudes, and why they hold on to outdated ideas. People unconsciously look to their tribe, or main identity group, for guidance, and are even more susceptible in times of uncertainty and change.

Allport didn't buy it. He disagreed with the premise that people's behaviour was shaped primarily by their membership of groups. For him, the individual's personality and personal experiences were key. To overcome a social problem such as racial prejudice, you looked to character building and introspection for answers.

Yet Sherif was impatient with and increasingly dismissive of the individualistic emphasis of North American social psychology. Psychology that was relevant and pertinent to building a new nation had to concern itself not with individual introspection but with understanding how citizens embraced the often faulty thinking and attitudes of those around them. The only true social psychology was one that studied people in groups.

Disheartened by Allport's reaction, Sherif quit Harvard soon afterwards, leaving for Columbia University.

Despite their intellectual falling out, Allport wrote a letter to Columbia introducing his former student, describing Sherif's work as 'unusually original' and Sherif as 'highly ambitious', with a complete 'new framework for social psychology' in mind. But in another letter to a friend in Turkey, Allport wrote glumly, 'I do hope he leaves his temperament behind him when he enters Columbia.'

Sherif loved New York. 'The whole world is here!' his daughter Joan remembers him saying. He revelled in its cosmopolitan character and multicultural mix, and walked the streets for hours each day.

He moved into a six-storey apartment block at 740 Riverside Drive in Harlem in April 1934. But he had no money and no obvious way of paying the rent in Apartment 3C. For a while, his new supervisor, Gardner Murphy, helped him out financially, even though Murphy could ill afford it. Yet once again Sherif was rescued, this time by the American Friends of Turkey, a charity dedicated to helping to rebuild and redevelop the country after the war of independence, whose founder, Asa Jennings, thought Sherif's work was brilliant. 'Sherif … was over to our apartment for dinner a few nights ago. We had a very splendid evening's visit with him. He has made a marvellous record over here and there is no doubt that he will go far. I am very anxious to see his book: it is evidently quite remarkable.'

The book was *The Psychology of Social Norms*, which Sherif began writing soon after he arrived with the help of his new sweetheart, Mary Alice Eaton, a student at Wellesley College who he met at Harvard. Sherif could be charming and was attractive to women, but they had to be prepared to put up with his work ethic. For a while he considered marrying Mary Alice, but decided against it because marriage 'might interfere with my developing work'.

Sherif's move to New York had been impulsive, kind of crazy. He'd lost his Harvard scholarship. It was the middle of the Depression, and he had little means of supporting himself. But in a way, it was the perfect environment for the research that he had planned for his PhD. The darkened room at Columbia, with its supposedly moving pinpoint of light, echoed the ambiguous and shifting landscape of social change on the streets outside. Sherif argued that his autokinetic experiment demonstrated how in periods of crisis — when the usual boundaries are blurred or unclear, and people feel uncertain — members of a group are more likely to be influenced by and accept the perceptions of others. It might not sound like much of a trailblazing experiment, but it made a splash. It seemed to offer tangible evidence of how social norms arise and are kept alive, as well as sounding a warning about the human tendency to be shaped and moulded by those around us, especially in periods of rapid social change.

And he had evidence of that right on his doorstep. Harlem was a magnet for intellectuals and artists, farm workers and labourers, where black Americans were finding a new voice and demanding greater racial equality. On streets and in churches and clubs and houses across Harlem, African Americans were forging a new cultural identity, spurred by the black soldiers returning from the war. A new generation were rejecting cultural stereotypes and envisioning a new future.

'Father Divine is Dean of the Universe.' The silver letters swayed from a line across the ceiling in the hall in Harlem. In the packed benches below, hundreds of people sang fervently, accompanied by a brass band:

*Father Divine is the captain,*
*Coming around the bend,*
*And the steering wheel's in his hand …*

The hall was hot, and handkerchiefs fluttered as people in the crowd mopped their faces. A large, middle-aged woman with a hat perched sideways stood and told the audience she'd suffered terrible pain and misery from her bad knee — no doctor could help her. But then she met Father Divine and he miraculously cured her. Some people listened quietly; others closed their eyes and moaned. Some shouted, 'We thank you, Father!' and 'Isn't he wonderful?'

The band started up again, and the woman led the next hymn, with people clapping and calling the praises of Father Divine. One by one, a continuous stream of people stood up to testify to his amazing power to reverse misfortune and transform lives. Downstairs, an enormous banquet was laid out, the plates continuously replenished for the constant stream of hungry disciples.

Muzafer Sherif spent most of his Sundays in 1934 here, at Father Divine's Revival meetings, watching, fascinated, at this proof that such a large and growing number of people could be convinced that a rather ordinary-looking man was God on earth. How and why had Father Divine managed to inspire such devotion? What psychological factors were at play that bound his followers together? Disguised as a believer, Sherif mingled with the crowd, surreptitiously recording proceedings and conversations with followers.

Sherif and fellow researcher Hadley Cantril noticed how the preacher created a self-contained world. In the realm of Father Divine, member-believers were promised not just material comfort but also pride, freedom from oppression, security, respect, belonging, racial equality, and certainty in a frightening and uncertain world. His followers traded their former individual identity for a new collective one, bound together by a common set of standards that gave them a sense of belonging.

It was in the Church of Father Divine that Sherif too saw the light, the clearest outline yet of his emerging theory of the social psychology of groups.

A charismatic leader inspires, protects, and gains cooperation through powerful rhetoric, but most of all by appealing to people's need to belong.

Sherif's postgraduate training might have been American, but his early studies were exploring a theme that was central for the modernisation of Turkey. Sherif was never explicit about this goal, or that Atatürk and the new Republic were the drive for his developing a new social psychology. But his country was experiencing a historical moment, when people who saw themselves as Ottoman subjects, their lives governed by the rituals and traditions of religion, were required to renounce that identity in favour of citizenship of a new secular Republic. It seemed no coincidence that Sherif's work mirrored the same question: how could they be persuaded to adopt new ways of thinking and behaving and embrace their new identity?

As Muzafer Sherif conducted his research and made his observations of the men's changing viewpoints on the pinpoint of light in the darkened room and of the congregation of Father Divine, he too was being shaped by forces around him.

He roamed New York's streets alternately inspired, appalled, thrilled, and outraged by what he saw. The country was in the grip of the Depression, and he was shocked to see the destitution and suffering, the growing number of people sleeping in parks, families camped in cardboard shelters, the lines for soup kitchens and charities stretching for blocks. Others walked the streets looking shell-shocked, or stood listening to soapbox orators urging them to join marches and protests, rallies and rent strikes. But he was exhilarated by the impassioned speeches at mass gatherings, the defiance and indignation that fuelled union membership, strikes, and protests, and the way people were organising and banding together to fight injustice at home and fascism abroad. He wrote to Gardner Murphy highlighting the irrelevance of religion at such a time:

> Last night I attended two meetings at Columbus Circle … one of the Salvation Army; the others called themselves hobo group … The speakers were attacking each other. The Salvation Army woman was preaching big words that are the heritage of centuries. The hobo speaker was pointing out six hungry men and saying, "We don't want pie in the sky. We want to feed these hungry men (with the emphasis of his fist going down) and now!"

But it was the communist rallies that had the biggest effect on him. One night he attended a rally of 15,000 people to protest the arrest of twenty-year-old communist and African American labour organiser Angelo Herndon, who had been sentenced to twenty years on a chain gang for 'inciting insurrection' among black workers in Georgia. Herndon and others recounted the legacies of slavery — the devastating effects of discrimination, segregation, and lynchings, and how the Great Depression had hit black Americans much harder than whites. In the heaving crowd, Sherif was electrified by the injustices Herndon described, and that night wrote to his supervisor that he felt outraged and embittered by what he'd heard. Sherif, like so many of his colleagues, turned to the communist party, which championed public infrastructure projects that employed unemployed people, agitated for racial equality and workers' rights, and opposed fascism overseas and oppressive economic practices at home.

On the streets of New York, Sherif witnessed what he saw as the cruel consequences of a capitalist society and the emergence of a collectivist identity. At soapboxes, street-corner meetings, and demonstrations, he watched the birth of new groups and saw how the disempowered became empowered, the downtrodden became defiant, the oppressed raise their voices, as a result of being part of a larger movement. By observing the speakers who, in their passionate commentary, described individual experiences that everyone in the audience could identify with, by dramatic gestures and rhetoric, they created a feeling of we-ness: a sense of equality, of shared identity that gave people solace and solidarity, and encouraged people to take action to look after one another.

At Columbia University, Sherif found his own tribe, joining a small group of psychologists from ethnic and racial minorities who identified strongly with the issue of racial prejudice. Among his friends were African American researchers Kenneth and Mamie Clark, whose work on the psychological effects of prejudice on black children later contributed to the US Supreme Court's ruling in the case of Brown v. Board of Education that racial segregation was unconstitutional.

It's ironic that the discipline that helped legitimise racism in the first two decades of the twentieth century was, two decades later, trying to dismantle its effects. By now, psychology was abandoning its efforts to prove differences in intelligence between ethnic groups to justify discrimination, to focus instead on the psychological causes of prejudice and how to overcome it. Social psychologists began to view prejudice — whether racial, national, political, economic, or class-based — as a set of harmful attitudes that led to injustice, mistreatment, feuds, bloodshed, and war.

Sherif and his colleagues were part of a growing number of intellectuals who shifted to the political left during the

Depression. They saw capitalism as perpetuating inequalities of opportunity and promoting a culture of discrimination, and dreamed of a society free from oppression, promising equality for all.

But as Sherif soon discovered on his return to Turkey in 1937, exposure to new ideas in America had changed the way he viewed his homeland. And it would transform him into an outcast.

In 1937, Sherif returned to Turkey to take a job at the Gazi Institute in Ankara and, two years later, transferred to the newly established Ankara University. Like the University of Istanbul, from which Sherif had graduated, the capital's new university had been established to further the Republic's revolutionary ideology and turn out graduates who were ardent supporters of the politics of modernisation.

In the time he had been away in America, Ankara had developed into a modern European city, although it had little of the beauty of Istanbul. Planned, orderly, regulated, it represented Atatürk's vision of the future and reflected the influence of the German architects who designed it. The rather sleepy town had been transformed, with roadways, wide boulevards, and public buildings in Bauhaus, Art Deco, and Cubist styles. Money and people flowed into the city — government officials, diplomats, teachers, students, businessmen, and shopkeepers. Ankara had few mosques, plenty of health centres and sports arenas, some cinemas and schools. But compared to Istanbul, there was little night-life. Underneath its modern exterior and despite being the country's capital, Ankara was still a village. And like any small village, it was rife with spies and gossip and intrigue.

'For everything in the world — for civilisation, for life, for success — the truest guide is knowledge and science.' Atatürk's positivist message was inscribed as a daily reminder to staff and students passing through the doors of the imposing Germanic façade of the Faculty of Languages, History, and Geography at the University of Ankara. I imagined that at first Muzafer Sherif hurried past the quotation on the wall with barely a second glance on his way into work each day. But he soon began to question exactly what kind of knowledge and science should guide the new nation.

Sherif, like many of his peers, had regarded Atatürk as a contemporary hero, with a vision for a new Turkey as a modern democracy: a state that could straddle both the secular and the religious, and implement the progressive reforms required to transform it into a nation equal to other modern nation-states in the world. Until his death in 1938, Atatürk kept extreme forms of nationalism under control and kept his distance from Nazi Germany. But after his death, the pro-fascists in the ruling party became more vocal and influential. Turkey's relationship and alignment with the ideology of Nazi Germany became increasingly close. Worse, the government was looking to import a capitalist model for the economy, and was more and more framing discourse about national identity according to race.

In America, Sherif had embraced the notion of social psychology with a conscience, that should and could document the psychological costs of oppression and inequality and help to dismantle oppressive political systems. And he'd been surrounded at Columbia by people who shared his idealism. But the Ankara faculty that he joined taught theories of hereditary racial inferiority and superiority as well as the new Turkish history, rewritten to prove the historical superiority of the Turkish 'race'. The faculty was a microcosm of European politics, divided into those, like the German or German-trained Kemalist professors, who supported such racist theories, and those, like Sherif and his friends Behice Boran and Niyazi Berkes, who had studied in the United States and brought criticisms of racist science back with them.

During World War II, Turkey aligned itself with whoever looked likely to win, signing friendship agreements first with France and Britain in 1939, then with Germany in 1941. The government's attitude to its leading intellectuals flip-flopped between tolerance and persecution. Until 1941, the government was supportive of journalists, publications, and intellectuals that favoured Britain, France and their Allies.

Sherif made no secret of his views about the war, and was 'violently outspoken' about his support for the Allies, and vociferous in his opposition to the axis powers. In a series of articles in left-wing journals, Sherif criticised those in Turkey who were sympathetic to the Nazi doctrine. He scoffed at supposedly scientific theories about race differences based on IQ or skull size. He called proponents of hereditary differences 'defectives' and labelled fascism as ignorant, feudal, and primitive, echoing the terminology the Young Turks used to describe the Ottoman Empire. Racism, Sherif argued, was incompatible with the modern, civilised Republic. He had arrived back in Turkey from America sure of his ideological high ground, and his ridiculing tone and dismissiveness infuriated his critics.

But between 1941 and 1943, with a German victory looking

likely, the government increased their efforts to demonstrate they were pro-German. Turanists who had adopted Nazi ideology called on the government to crack down on anti-fascist left-wing intellectuals and writers. In 1943, when Sherif published *Race Psychology* (*Irk Psikolojisi*), a book of essays based on the lectures of Canadian psychologist Otto Klineberg that debunked the idea of racial inferiority, antagonism between the right and left in Turkey was intense.

Government foreign policy shifted again in 1943 towards Britain, France, and Russia, with the defeat at Stalingrad. It became clear Germany was losing the war.

In a biography of Niyazi Berkes, a fellow student of Sherif's in Istanbul and by this point colleague and fellow left-wing professor at Ankara, the author Şakir Dinçşahin suggests that the personal friction between Berkes and Sherif sprang from Sherif's early politics and his former enthusiasm for the racist ideology he now so bitterly opposed. He wrote that Sherif 'had been under the influence of the ultranationalist ideologies when he was an undergraduate student … but changed his politics in the course of his graduate studies at Columbia University and became a leftist intellectual'.

I hadn't come across the suggestion before, and it made me pause. Was Sherif's opposition to racism in the 1940s a case of the zealotry of a new convert, or was it motivated by something deeper? I wondered if his interest in the power of groups to shape individual attitudes and judgements sprang from his own experience embracing attitudes and ideologies he later regretted? Was the Robbers Cave experiment an attempt to understand how he himself had been swayed by a group to do or say things he was ashamed of? Or, alternatively, was he using his own research to show how he had discarded his own beliefs in racist ideology, applying science to himself? That seemed gutsy to me if so.

He certainly sounded unusually provocative in his views. Even his left-wing friends felt Sherif's goading went too far. During a railway tour of Anatolia to promote Ankara University in 1944, a prominent military veteran held forth to a crowd on the superiority of the Turkish race. In Turkey, the military enjoyed a privileged place in society particularly for their role in the War of Independence, and public criticism of them was taboo. As proof of the superiority and longevity of the Turkish race, the veteran told a crowd that the pastirma for which Kayseri Province was famed had been brought there in ancient times by the warrior tribes of Central Asia, the forebears of modern Turks. Sherif heckled him: 'The people of Kayseri will erect a statue of you,' he said mockingly. 'They'll make it from pastirma.' Insulting a military officer in such a public way seemed particularly dangerous. Police later questioned Sherif, but no charges were laid.

Meanwhile, Sherif continued his attacks in a leading left-wing magazine, calling for the redistribution of wealth and power in Turkey and antagonising the old-guard ruling class — which included landowners, educators, government officials, and even members of his own family. Many of these individuals 'hated and feared' him, Carroll Pratt, an American faculty member who worked at Ankara University soon after Sherif, later recalled. The intellectual elite in Turkey was small, and such public criticism was hard to ignore. Sherif made many enemies among the influential in Turkey, and he would find that some held grudges against him for years afterwards.

After the tide of war began to turn against Germany in 1943, Marxist scholars, of which Sherif was considered one, became a target of attack by members of the far right. Prominent pan-Turkists accused the government of protecting communists and described a number of professors at the university as 'traitors to the fatherland'. On 12 February 1944, after a student at the military academy was caught with Communist Party propaganda, the government instigated an immediate crackdown and began a mass arrest of prominent leftist intellectuals.

Despite being tipped off that his arrest was likely, Sherif seemed to think he was indomitable. Unlike his friends, he made no attempt to hide and, according to his daughter Sue, was surprised when he was arrested as he waited at a bus stop one morning on his way to the university. It's hard to imagine how he carried this sense of immunity given the political atmosphere at the time: how he thought his independent and outspoken views would be tolerated.

Sherif was accused of promoting Bolshevism and having inappropriate relationships with female students at the university — whether this latter charge was concocted or had some basis in fact it's impossible to tell. One of Sherif's Oklahoma students recounted how Sherif had told them he had been victimised for stepping in to protect a female graduate student who was going to be dropped from the university's graduate program 'for no other reason than being a Jew'. And that Sherif, it was said, 'warned the other professors that if they failed her he would fail all their other students'. But unlike his friends and colleagues who were taken to jail, tortured, and forced to stand trial, he was released after four weeks of detention in a former school. It was his powerful connections that saved him: his brother was a prominent member and supporter of the ruling Republican People's Party and a close friend of the prime minister, who also came from Ödemiş. Sherif had a private meeting with the prime minister, who promised to keep him from trial if Sherif gave his word that he would leave the country.

After his release, Sherif was deeply depressed. His Communist Party friends felt betrayed and angry, and closed ranks against him. He was isolated, and suspicious of university colleagues who he thought had spied on him. He applied for a leave of absence from the university, and arranged for his former Harvard teacher, Carroll Pratt, to temporarily fill his position. He wrote to Gordon Allport, telling him that he had 'failed' in convincing anyone in Turkey that he had 'anything to contribute' as a social psychologist and asking Allport to help him find work in America.

Eight months later, in January 1945, Sherif boarded a plane for the United States. But the chill wind of the Cold War would soon be blowing in America, and it would not be a welcoming place for people with his political past.

# 13
# Oklahoma

Sherif said when he arrived in America that he felt like Rip van Winkle, a fairytale character who goes to sleep and wakes up to find twenty years have passed and the world has completely changed. He had been cut off from America during the war years, and when he arrived back one of the biggest changes he must have noticed was the disappearance of the vibrant left-wing culture he had experienced in the 1930s — the coalition of causes, unions, and groups pushing for social change had disintegrated.

Sherif arrived in Washington in the winter of 1945 and was met by his Harvard friend Hadley Cantril, who had used his connections to get Sherif a two-year fellowship at Princeton, and proposed they write a book together. Sherif was still low in spirits, and progress on the book was slow.

When a social psychology graduate called Carolyn Wood wrote to Cantril in October 1945 asking about a job, he invited her to Princeton to talk about working as a research assistant to Sherif. In his diary after his first meeting with her, Sherif congratulated himself on not immediately asking her out to dinner.

Like Sherif, Carolyn Wood was an achiever, ambitious, and idealistic. She was a straight-A student of social psychology, a discipline she fell in love with after reading Sherif's first book, *The Psychology of Social Norms*, admiring 'its beauty and logic'. As well as being highly intelligent, she was Hollywood-starlet good-looking, musical, and outgoing. At college, she hosted a radio show, acted on stage, and sang in a quartet. Within a month of their meeting, Sherif described Carolyn as 'the center of my universe'. By December, less than two months after meeting, they married. She was twenty-three, he was forty.

Sherif wrote draft after draft of a letter to Carolyn's parents, attempting to reassure them about a marriage that must have seemed sudden and impulsive:

> I understand fully your deep concern about her … Many times … I put myself in your situation facing this sudden marriage of ours which naturally appears queer and appalling to you … It would be very understandable that you should think she is impetuously carried away by a fascination of [sic] things strange and foreign and that someday she will wake up to realize what a damn fool she has been to fall into such a trap. I call her attention to this possibility … I realize quite well that nothing I write or say now will eliminate your apprehensions and consternation …

Any woman Sherif married would have to be as devoted to social psychology as he was. Carolyn Wood would be just as much a scientific partner as a lover and a wife. He wrote to a friend to tell him he was 'deeply in love', reassuring him that

Carolyn was an excellent match: as well as being 'a mid Western beauty … she'll share all of our enthusiasm [for social psychology] … She'll add new sparkle to it.' The two made plans about their future life in Turkey, where Carolyn would be able to pursue her career and teach at the university. Announcing the wedding to Gardner Murphy, Sherif said, 'I did my very best not to give her any rosy picture about my future. In fact, I was grimly realistic.' It's not clear exactly what he was so 'grim' about, or why the letter, rather than brimming with joy or anticipation, is weighed down by melodramatic foreboding: 'From now on, I am basing all my work, both theoretical and ideological, on my relationship with her. I shall pull through or fail utterly in my work and everything on the basis of this relationship …'

Carolyn too was looking for an equal relationship. Sherif's reputation and his 'fervour' for male–female equality immediately attracted her. 'I wanted to marry an intellectual, as well as a sexual and emotional partner, who would encourage me being a social psychologist,' she wrote in an article about how she came to her choice of career. Muzafer Sherif seemed to fit the bill. But she would get more than she bargained for, because as Sherif's letter hinted to Gardner Murphy, he would be dependent on Carolyn to balance his 'craziness' as well as being his colleague and partner.

In theory, Sherif's new research assistant was meant to free him up from research so he could focus on writing his share of the book he and Cantril were co-authoring. But Cantril, frustrated, pressured Sherif to hurry and finish. Sherif did not take it well, writing to Cantril that such a book shouldn't be hurried and that 'in order not to strain our relationship and my stay in Princeton further', he would work separately and at home.

I don't know if Cantril spelled it out, but he was uncomfortable with Sherif's flaunting of pro-communist views in his writing. During the war, Cantril had worked for the Office of Strategic Services (the predecessor of the Central Intelligence Agency) and developed strong ties in government. Now the ever-entrepreneurial academic was positioning himself to win government funding to work on anti-communist psychological warfare. Like many of Sherif's colleagues, Cantril, who had been a member of a number of thriving left-wing groups a decade earlier, was now careful to downplay his pre-war political affiliations and activism.

I wondered how much of this tension with Cantril was a spillover of Sherif's worry about Turkey. He was still agitated and upset by events there, and followed the appointment of replacement faculty at Ankara closely, writing letters home describing the new appointees as 'SOBs, fascists, Nazis, spies and blackmailers'. Despite his loathing of many of these colleagues, in January 1947, with his Princeton fellowship over and Carolyn pregnant with their first child, the couple continued with plans to return to Turkey later that year. He would have known from reports in the American press that the Turkish government was engaged in a 'Red hunt', and in particular was intent on ridding higher education of communists and subversives. Yet he seemed genuinely shocked in May 1947 at news that he had been sacked from his job at the university in Ankara, ostensibly because he had broken a Turkish law that banned anyone married to a foreigner from working as a civil servant. But it was more likely part of a wider crackdown as Turkey proved to its new Cold War ally, America, that it was taking its responsibilities seriously and dealing firmly with its left-wing intellectuals.

It was a blow for Carolyn's academic ambitions too, as the plans for her to work in Turkey evaporated. With the influx of soldiers returning after the war, opportunities for women in American academia were shrinking fast. At the same time, Sherif's American visa had expired and unless the state department agreed to renew it, he looked like being deported. Instead of returning home with his wife and child to a relatively comfortable life in Turkey, Sherif found himself stranded, one of a wave of immigrant psychologists who had arrived in America after the war. Carolyn comforted him, writing to him when he was away visiting Yale, 'My sweetheart … as long as we are in good health we'll be able to take as much as stupid people can give — and can find ways of maintaining ourselves. I am with you every step of the way.'

Luckily, Yale psychologist Carl Hovland was recruiting scholars to work with him and organised a two-year Rockefeller Fellowship for Sherif. It would give him some breathing space while Carolyn tried to sort out his immigration problems.

At thirty, Carl Hovland was ten years younger than Sherif but he looked older than his years. In the portrait at Yale taken around this time, his wavy hair was already showing signs of grey. His solid body and the avuncular pose suggested someone calm and unflappable. Hovland was careful, methodical, smart; he was the antithesis of Sherif, who had big ideas and an excitable personality. But Sherif had been cut off from America and its developments in psychological research during the war, so there was a lot of catching up to do as he cast around for a new research project.

Sherif's immigrations woes continued. He didn't help matters. Impatient with anything that smacked of bureaucracy, he often simply ignored official-looking letters and did not return phone calls that looked like they might have something to do with red tape. Throughout 1947, he skipped appointments with the immigration department, telling them he had important business to attend to at Yale. And he'd missed letters warning about the expiry of his temporary visa.

Meanwhile, any chance of him being welcome back in Turkey was diminishing fast. At the University of Ankara, things were getting worse. Right-wing faculty and students protested against leftist teachers. Student demonstrators broke into the offices of the university's president, demanding the resignation of three of Sherif's friends, who were arrested and put on trial. The persecution of intellectuals seen as sympathetic to communism in Turkey continued with the murder of prominent left-wing writer Sabahattin Ali by Turkish security services. If Sherif applied for a new visa, the American government was sure to ask the Turkish authorities for a report on his politics and his allegiances: his avowed sympathy for communism was sure to come to light. In July 1949, at the time of the first summer-camp experiment, he was in the country as an illegal alien. With all this hanging over him, Sherif threw himself into preparations for his research.

Once the first summer-camp experiment was over, Sherif drafted a report for Hovland. In this first study, the twenty-four boys who began as friends turned on one another as enemies, after a competition for prizes. But there are hints that the process did not go well. Sherif was dismayed to find the camp had no electricity, so there was no way of gathering recordings, and in a letter to Hovland he hinted at a lack of staff cooperation. And while Carolyn admired the study, she referred to it in later years as 'a mess' and noted how the data gained from it could not be used.

Whatever happened, it had exacted a toll on Sherif. He seemed to disappear once it was over. The American Jewish Committee pleaded with him for six months to provide them with details of how the experiment had turned out. In one letter, the AJC's Joseph Flowerman wrote that he had two theories about Sherif's silence. Either he was 'head over heels analysing his data' or was 'too unhappy to talk about it'. Clearly exasperated, he ended the letter asking for evidence Sherif was still alive. Sherif replied almost immediately, saying the letter had been like 'shock treatment', and that his 'seclusion' was due to 'complete exhaustion'.

On top of all this, he explained, he had moved to Oklahoma. You can feel his impatience at the disruption of the move to Oklahoma at a time when he 'was so keyed up'. Moving house, settling in, setting up an office — all these interruptions 'made me more and more restless and caused me to lose my contact with everything else'. But now he was over the moon with the results, he told Flowerman:

> This study has been the most exciting research unit for me as a social psychologist with all the background I have in the field. It has been a revelation in concrete form of what I have been thinking theoretically for several years … To use your characterization of last year, my 'libido' has been and continues to be directed towards this study.

What he didn't say was that he had already written a report

but it had been through several drafts on Hovland's advice.

In his first draft, Sherif concluded that in this study, the boys' behaviour reflected the dynamics of a competitive society that divided people into the 'haves and have-nots', stoked rivalry and resentment, and fostered prejudices and, eventually, violence. The report echoed the themes of both of Sherif's books, including the one co-written with Hadley Cantril, in which he had expressed sympathy for Marxism, and admiration for what he saw as the benefits of Russian collectivism over American individualistic and capitalist culture.

A wave of anti-communism was sweeping across America in the late 1940s. Unlike so many of his peers, Sherif seemed indifferent or unaware that political conservatives would view such research as suspiciously pro-communist. It was a dangerous time for a psychological scientist to be seen to be anything but impartial. Especially one without a visa, whose status in the United States was still so uncertain. But he seemed oblivious to the peril of expressing such a view.

Between the early draft of a paper about the study that he sent to Hovland in November 1949 and its publication in April 1950, China fell to the communists, the Soviets detonated the atom bomb, and Senator Joe McCarthy made the sensational announcement that he had a list of the names of communist spies in the state department.

Hovland suggested changes to the draft. Given that Hovland provided high-level advice to a number of leading organisations who funded communications research, including the US Air Force, he was well placed to advise Sherif. Hovland — the precise, careful experimenter — was also on the board of several military-related funding bodies and knew both how to frame research findings to suit an audience and how important it was to appear neutral and systematic. In his own research, Hovland kept his focus narrow and was very careful never to show his hand when it came to ideology or beliefs.

This wasn't unusual. Small-group research, which had thrived during the pre-war years, was adapted for the military, so social scientists, and particularly those with a political-activist past, were careful to reframe their projects as politically neutral. For example, those who promoted racial equality and argued that systemic injustices such as poor education, poor nutrition, and poverty disproportionately affected minorities reframed their stance away from notions of changing the balance of political power. Researchers in racial prejudice keen to distance themselves from leftist politics shifted away from social revolution to politically safer territory. Instead, they focused on just one part of the solution — altering people's attitudes and thinking, which had none of the connotations of social revolution. In a time when the boundaries of what was politically acceptable shrank daily, appearing as an impartial and value-free scientist with no political allegiances and no commitment to an ideology offered some protection against attack.

Sherif took Hovland's advice to heart. In the new draft, a kind of paralysis overtakes his writing. Gone are references to class, how the experiment reflected the dynamics of a capitalist society, or the alienation the system breeds between workers who regard one another as rivals in an economic competition. Any inference that a capitalist system sets up inequality between groups in society by granting unequal access to money, power, or resources, and so breeds social discord, was gone. Sherif's language in the final draft was sanitised, cleansed — and deadly dull. There was no reference to real-world politics. There was no longer anything revolutionary lurking in those pages.

The final published version of Sherif's 1949 study was scrupulously anchored in the detail of the campsite at Happy Valley and made no reference to a broader social or political context. Somewhere between the earlier drafts, his recklessness had been replaced with caution. Any mess or uncertainty was replaced with confidence and fact.

Sherif downplayed issues that might raise questions about the study's ethics and removed reminders that the subjects were children. Between the drafts and the published report, the word 'games' was substituted with 'competitions', 'adults' became 'experimenters', and 'children' became 'subjects'. Between the drafts, staff became more rigorous. In an early draft, a participant observer had the 'assistance of the Junior Counsellor who was under his supervision', but this was later reworded; the junior counsellor in the published version of the report was 'under [the participant observer's] direct control and followed his lead'. Tellingly, Sherif wrote that staff were under strict instructions to follow the experimental protocols, hinting that there were lapses: 'The tendency to depart from the observance of these instructions on the part of any staff member was forcefully called to his attention so that it might be corrected.' The actions of staff in generating resentment and conflict among the children, while it was mentioned, was downplayed as 'mildly frustrating situations'.

Of course, I expected a draft to be transformed by the time it appeared in print, but the final description of the two groups of boys reads like a caricature. Named the Bull Dogs and the Red Devils, they could as easily have been called the good guys and the bad guys. In the Bull Dogs group, one boy called Crandall rose to leader by getting others on board in planning and carrying out activities, such as putting the letter 'B' on the door of the bunkhouse and building a 'chinning bar'. He also led the group on a hike, and was helpful to others by bandaging a blister and fixing a belt. He used praise and encouraged the others in the group to do the same: 'We did a good job, boys. We should be proud of ourselves.'

If the Bull Dogs were a model of the democratic group, the Red Devils were the opposite. Their leader was Adam 'Babyface' Severin, a tall, good-looking boy who looked fifteen instead of eleven, and whose nickname conjures images of gangsters and thuggery. Severin became leader through his 'daring' and 'toughness': Sherif described Severin as tyrannical and autocratic in his dealings with the other members of his group, all of whom called him 'Captain'. He 'enforced his decisions by threats and actual physical encounters' on members of his own group. When two or three of these boys were seen to be seeking out their old friends — now members of the rival Bull Dogs — Severin branded them as 'traitors' and his henchmen 'threatened them with beatings'.

Sherif's description of the differences between the two groups reads just like a description of the differences between democratic and autocratic groups Kurt Lewin and his colleagues had described ten years earlier. The symbolism of their names, too, seems no coincidence. The victors and winners of the knives, the Bull Dogs — named after Yale's mascot — have a fair-minded leader who inspired his group and fostered an atmosphere of mutual respect and encouragement. The dysfunctional Red Devils, with their autocratic 'captain' and his 'lieutenants', using bullying tactics to subjugate other members of the group, are the losers in the competition. It seemed just that it was the Red Devils who lost the competition, and inevitable that they were the treacherous ones, planning sneak attacks when the other group thought 'war' was over.

It's impossible not to read Sherif's description of the two groups and their fate as a tribute to the superiority of democracy over totalitarianism. Even before his account of the study had been published, it caught the attention of American military, who saw a range of useful applications for research about cohesion and relationships between small groups. Whether he was aware of this possibility or not, Sherif's results could be used in ways he may never have considered. Getting black and white, American and foreign troops, working together was critical in wartime. Without esprit de corps, troops were likely to surrender, defect, or lose the battle. Small-group research could also inform efforts for using peer pressure to get reluctant soldiers enthusiastic about fighting, and influencing how citizens would embrace or reject propaganda.

Yet when members of the military first approached Sherif and expressed their interest in funding more of his research, he wrote to Hovland, telling him the news, and seemed genuinely surprised.

Over the next decade, Sherif worked on several projects funded by the military, and apparently turned down some that involved studying the effects of being in a fallout shelter because he was opposed to the nuclear arms race. So how was it that such a left-wing psychologist, interested in group conflict with the aim of engineering peace, had not been alert to the broader political and social currents that circulated around his work and the potential ways his results could be used for political and military advantage?

At the University of Oklahoma campus, inspired by the architecture of Oxford and Cambridge, the turrets of the gothic buildings rise above the trees, and the curving archways and leadlight windows create a sense of history and tranquillity. Sherif had been wooed to the university, which was looking to improve its standing. His international reputation and ability to garner research funds brought prestige and status to a university that was more famous for its football team than its academic credentials. When Sherif arrived there, the university was little more than fifty years old, and underneath its peaceful veneer Cold War tensions crackled.

His first four years at the university were nightmarish at times. In March 1951, the Oklahoma state government turned its attention to local universities as part of its hunt for communists. He and Carolyn built a fire and burned incriminating books and papers. The government passed a law requiring all local and state employees to sign a loyalty oath. Sherif signed the oath that same month, swearing that he was not a communist, that he would be loyal to America, and that, in the event of an attack, he would bear arms and fight for the nation. His daughters remembered the tension and anxiety of those McCarthy years and how one afternoon, their parents, unexpectedly dressed up, with their father in a suit and tie and their mother in her best dress, told them to play quietly in their room while a group of men in suits came to visit.

In April 1951, Sherif was arrested because his visa had again expired. The US government had plans to deport him. Turkey was an increasingly dangerous place for leftist scholars, and Carolyn lobbied furiously on his behalf, marshalling all her contacts and acquaintances, arguing that his pro-Allied stance in Turkey at a time when such views were unfashionable proved his loyalty to America. That same year, the Turkish government instigated proceedings against him to recoup the salary he'd been paid during his leave of absence. His brother, the lawyer, represented Sherif at the trial, but their relationship soured as a result. Another major communist trial was also underway in Turkey, and Sherif stopped writing letters home around this time, probably because many of his friends were implicated. He became circumspect about talking about Turkey in public. At a speech he gave at Columbia during this period, he told the audience he had seen 'human conflict and misery of all kinds in one of the very delicate spots of the world — my own country of Turkey. I cannot go into concrete details as I would like to tonight, but prefer to stay on the theoretical level. I hope the implications will be clear.'

The continued military support for and interest in his research did not assuage government concerns about his patriotism or loyalties. Soon after arriving at the University of Oklahoma, Sherif began work on research for the navy using his autokinetic experiment in officer selection training. As a foreign national working on US Navy projects, Sherif came under suspicion. In August 1952, J Edgar Hoover directed the Oklahoma branch of the FBI to investigate him. The investigation lasted almost a year, and aimed to establish whether Sherif was a security risk and whether he had access to any classified information as part of his work on naval research. An FBI investigation into communism among Turkish students in America also turned up his name. FBI agents across the country interviewed Sherif's mentors and colleagues, librarians and landladies, shopkeepers and administrative staff who had known him in America, as well as some who had known him in Turkey, about their views of his communist leanings. At the same time, many members of the Society for the Study of Psychological Issues — an organisation of psychologists committed to pacifism and social change, including his mentors Gordon Allport and Gardner Murphy — were also under investigation.

Did Sherif notice on those afternoons when he called into the campus post office to collect his mail that the postmistress could no longer meet his eye? Did he ever wonder who the stranger was lurking in the corridor outside the dean's office?

I'd like to think someone let him know about the FBI investigation. I'm not sure any of them did. But Sherif was clearly wary. OJ Harvey remembered that one night he was giving a review paper about classless societies, and a couple of men who were visiting from the American Psychological Association wanted to sit in on the class. 'Sherif was very concerned. He wasn't sure they were from the APA, he thought they were from the government. He was concerned because he was going to talk about communism in the context of my paper on classlessness but he was worried about whether we should talk about something else instead,' OJ recalled.

Most of the people interviewed by the FBI reported that Sherif was a respected and even exceptional scholar, but it's striking how many said they didn't like him. At first I wondered if this was a ploy, an attempt to distance themselves from someone who was under suspicion. But the more I read, the more obvious it became that Sherif invested little energy with his peers in being likeable. One colleague from Sherif's time at Princeton, Herbert Langfeld, said that he and Sherif talked mostly about psychology and never about politics, and 'although he did not especially like him due to his arrogance, conceit and a desire to be the center of attraction at all times', he had no reason to think either Sherif or his wife were communist or Russian sympathisers.

Sherif's by now former friend Hadley Cantril described Sherif as so thoroughly 'obnoxious' that he had arranged for Sherif to work from home instead of coming to campus. While Sherif was an 'outstanding' scholar, he was very 'unstable and erratic … arrogant, highly conceited, intellectually dishonest and a hypocrite'. Cantril said he suspected that Sherif was 'psychotic'. When they worked together at Princeton, Cantril said Sherif continually voiced his admiration of the Soviet Union and the way things were handled there. Cantril told the FBI he believed that Sherif 'would have no hesitation in providing all the information he might possess to the Russians'. Cantril likely knew better than most how tenuous Sherif's status was in America, and how dangerous things would be for him in Turkey. So I was intrigued that Sherif's former friend had turned on him this way.

I tried to get to the bottom of Cantril's accusation. I wrote to some scholars far more familiar with the man than me, but they could only speculate that perhaps it was a case of professional jealousy. Yet surely there had to be more to it than petty vindictiveness? Despite working for the government on anti-communist psychological warfare, Cantril himself was subject to extensive surveillance and repeated interrogations of his loyalty up until 1956. For example, his FBI files show that at one point Cantril was implicated in a major Soviet spy ring. Agents wrote that Cantril had been responsible for recommending an important Soviet spy for a government job, and his name was found in the address book of another spy accused of passing on material from the Office of Strategic Services to the Russians. In his interviews with the FBI, Cantril seems to have used the opportunity to feed them damaging information about colleagues such as Sherif, but whether it was as a way to eliminate competition, realise a grudge, or prove his own cooperativeness and loyalty, it was impossible to say.

No one else repeated Cantril's claim about Sherif and the Russians. But all repeated Cantril's complaint about how difficult Sherif could be to get along with. According to his colleagues at Oklahoma, he was 'not popular with either students or faculty' because of his overzealous focus on work. He spent 'every waking hour in pursuit of some psychological theory'. They implied that Sherif was given fewer teaching hours because of his unpopularity. Instead, he spent more of his time on research with graduate students. Another said that when Sherif first arrived, it was clear 'other professors would not be able to tolerate him because of personal characteristics', but there was no question he had an outstanding international reputation that was an asset to the university.

If the university was a comparative backwater, it did allow

someone as competitive as Sherif to feel important. He showed up in the office of the dean or the university president unannounced, ignoring the protocols of making appointments, walking straight past the desks of their secretaries. He convened meetings of colleagues from across the country in a series of symposiums, and publicly and tactlessly pointed out the weaknesses as well as the strengths of their presentations. For Sherif at this time, it seemed that the idea of cooperation over competition was all right in theory, but he had trouble putting it into practice. He often got on better with his graduate students than his colleagues, and with peers outside rather than inside his own discipline.

As a teacher, Sherif could be 'tough, energetic, demanding … ruthless and heavy-handed' at the same time he could be 'warm, affectionate, patient … and would go out of his way and to great lengths to help people who he thought needed and deserved his help'. But sometimes his assistance could be disconcerting, especially during the final academic step for graduate students, the oral examination, in which they defended their thesis to a university committee. It was not unusual if Sherif thought the student wasn't doing a good job for him to butt in and take over the defence for them.

But woe betide any student in Sherif's classes who admitted to admiring psychoanalysis or behaviourism, which Sherif dismissed so roundly. Social psychologist Serge Moscovici wrote in a review of one of Sherif's books that Sherif's scornful attitude to researchers working on similar problems was 'unjust'. Moscovici wondered: 'Why reserve all affection for humanity in general and so little for his colleagues in particular?'

But while his closest colleagues may have found his arrogance off-putting, his wife Carolyn remained his most steadfast fan. Soon after they arrived in Oklahoma, the local paper wrote a story about Muzafer, calling him 'one of the world's greatest psychologists'. But Carolyn complained to a friend that they'd got it wrong: 'He is *the* greatest.'

At first I read these descriptions of Sherif's complex and sometimes contradictory behaviour as something new, pointing, perhaps, to disappointment at trading the prestige of Yale for Oklahoma, or to the way in which his work had overtaken his life. But then I remembered letters between his mentors Gordon Allport and Beryl Parker twenty years earlier that hinted at difficulties with Sherif's temperament — his intelligence and intractability, his emotional highs and lows. Was it that in Oklahoma the stress of recent events were taking their toll on Sherif, providing an environment for his temperamental inclinations to flourish?

If you're lucky, you come across a teacher in your life who changes everything. A door swings open, a previously dark room is filled with light, and beyond the doorway a whole vivid world springs to life. OJ Harvey felt this way in his first class with Muzafer Sherif at the University of Oklahoma. Until then, psychology for twenty-four-year-old Harvey had been 'a puny little science'— simple, dull, and myopic in its focus on laboratory life. But Sherif's psychology incorporated notions from history, anthropology, and sociology. For Harvey, it was a whole new world.

The two men hit it off, largely, Harvey believes, because he corrected Sherif during a class — something his classmates were too afraid to do. Soon Harvey was Sherif's teaching assistant, then his right-hand man. Sherif was highly dependent on the practical people in his life. Carolyn ran the household, wrote the family letters, took care of the children, answered calls for Sherif, helped to draft and refine book chapters and papers, paid the bills, and smoothed ruffled feathers. At the university, Harvey played a similar role. Sherif's reputation for unpredictability meant that colleagues and students didn't know what sort of reception they'd get when they requested a meeting with him or asked for a response to an administrative request — whether they'd be greeted by the cordial and charming professor or his intimidating alter-ego, who was overbearing and argumentative. So they went to Harvey instead. Harvey was well known for his organisational skills — he told me his friends said he could 'organise a bucket of worms' — and his diplomacy.

They made a strange pair. On one side was the Turkish professor with a relatively privileged upbringing and clear sense of entitlement. On the other was the son of a sharecropper who had saved for college from the age of ten by fattening up and selling orphaned calves. But his students' achievements were Sherif's achievements, and he encouraged pupils like Harvey to aim for the Ivy League.

Few knew Harvey's secret to gaining Sherif's respect. In 1950, soon after he started working for Sherif, the two had a run-in over an administrative mix-up. Sherif shouted at Harvey and, among other things, called him an idiot. Harvey told Sherif, 'Screw you! Nobody talks to me like that!' and quit. After a week, Sherif sought Harvey out, hugged him, and cried. Harvey said that Sherif trusted him after that, and treated him like an equal, including him in seminars and conferences where he was the only student among leading scholars in the field.

Sherif expected his graduate students to work as hard as he did, and those that didn't soon fell by the wayside. His evening seminars, scheduled to finish at 9.00 pm, frequently continued well into the night, usually at the bar. If a student missed a class, he would go looking for them afterwards to publicly remonstrate with them about their absence. Some found his compulsive drive and single-mindedness repellent; others found it compelling. But social psychology was his life, and he expected his students to take it as seriously as he did. OJ Harvey shared his zeal, and it was this shared passion that allowed the two men to work so closely together on what became one of the most imaginative social psychology experiments of its time.

By May 1953, Sherif's troubles seemed to be over. After two years of intense lobbying by Carolyn, and with the help of an immigration lawyer, his deportation had been averted. Despite Cantril's damning testimony against Sherif, the FBI found no proof and closed the case against him. And while his position at the University of Oklahoma was untenured and dependent on the amount of research funding he could bring into the university, the funding for a more ambitious group experiment, at $38,000, was huge — equivalent to seven years of Sherif's salary.

Carolyn did her best to help him choose the right staff. For this 1953 study — the Middle Grove study — Sherif chose people such as OJ Harvey and Marvin Sussman, who shared his work ethic and would be grateful for the opportunity to work with him. But Sherif seemed unable to shake a feeling of dread and foreboding. Did he sense early on that this major study would end in disaster? Or that he would have to salvage what he could of his theory and his reputation in an unplanned and desperate final experiment down in Oklahoma a year later? His fear of failure haunted him. Unfortunately for Sherif, he was unable to take the one person who was his greatest asset to this major study with him. Carolyn, with her knack for anticipating and resolving problems and her ability to keep Sherif in line, would be staying behind to take care of their daughters, aged six and three. That would turn out to be his biggest mistake.

# 14
# The Museum of Innocence

In 1975, Muzafer Sherif wrote to OJ Harvey out of the blue. They had been largely out of touch for a number of years, beyond exchanging occasional cards at Christmas. But in 1975 Sherif, in an apparently buoyant frame of mind, wrote saying how 'terribly good' he felt about the Robbers Cave experiment and how 'rejuvenated' he was when he thought about it:

> The more I think about it … the more I realize how much [the experiment] owes to you. Without your utmost participation, your resourcefulness, sound judgments and down to earth arrangements it couldn't possibly be carried through to its culmination. I just wanted to convey to you this strong feeling and my gratitude to you.

It was the first time that Sherif had acknowledged this to Harvey, although in 1967, at a talk at the University of North Dakota, he described the Robbers Cave study as 'the crowning one, done where the best things are done, in Oklahoma by Indians'.

It was a rare time — perhaps the only time in Sherif's professional life when he ceded control. OJ thinks that Sherif wanted someone to place limits on him. 'You would have thought that after the 1953 study, when I threatened him with the block of wood, that there'd be problems.' OJ laughed and shook his head. 'But he seemed to want someone to take charge.'

Listening to OJ's account of how hard they worked at Robbers Cave and how it took weeks for him to get back into the habit of sleeping more than four hours a night, I got a glimpse of Sherif's drive and his ability to inspire those around him: 'All of us went for broke, we were so committed.' Sherif and OJ had a particular bond. 'It was a heart-and-soul thing for both of us. We put everything into it,' OJ said. I recalled how OJ had once told me that he and Sherif would have thrown themselves off a cliff if the study failed.

I was still surprised by how open OJ was in admitting that Sherif knew exactly what he wanted to prove at Robbers Cave and why it had been so important to have things worked out in advance. 'We knew that that kind of thing happened, and we just reproduced it. Looking back, I don't think we did find anything that was not known about groups and harmony and tension,' OJ said. He saw it as his job to engineer the unfolding of Sherif's theory in the landscape at Robbers Cave. This emphasis on confirmation rather than investigation had always intrigued me, right since I first began to research the experiment, and it wasn't until I did some more reading that I understood this was a tradition from Kurt Lewin: to use experiments as a way of showing 'the way things work in the real world', where your research was a confirmation of what you already intuitively knew.

On one hand, it was a refreshing contrast to the often gimmicky kind of laboratory-based social psychological research of the era that valued counterintuitive and surprising results. On the other hand, acting out a theory at Robbers Cave State Park and looking for confirmation of truth allowed for a casual cruelty towards the subjects in the study.

This apparent dichotomy that many social psychologists of the period seemed to share, between altruism and deep concern for humankind and an apparent lack of concern for the psychological wellbeing of their subjects, made sense to me now. I didn't agree with it, but I could see how Sherif and OJ could hold two such contradictory views at once. I understood now how Sherif was able to deny that there were any ethical problems with his research. After all, he saw social relationships as 'messy, contradictory and fraught with conflict, suffering and agony', and his responsibility as a kind of social engineer, intent on rectifying injustice and improving the world. The discomfort of a group of eleven-year-old boys was a small price to pay for research that could alleviate any of that torment. It explained why both Sherif and Carolyn saw social psychology as a vocation rather than just a career. And while they were forced to downplay this during the McCarthy years, their brand of idealism never left them.

OJ didn't agree with Sherif's conclusion that you could overcome conflict as easily as Sherif's experiment appeared to demonstrate. But he never confided this in Sherif. 'Afterwards, he wrote and told me it was the most important thing he ever did. It defined him. He said it was the biggest thing in his life. So I kept my opinions to myself.'

I could see what OJ meant. It might have been easy to

manipulate a peaceful resolution at Robbers Cave with groups of children, but how could you bring about a similar result in the broader world, when the gap between the haves and the have-nots was as wide as it had ever been, and the discrimination was systemic? But if you raised these kinds of questions with Sherif, OJ said, he shut the discussion down. There was no surer way of bringing an end to Sherif's evening seminars than by pointing out the degree of control the men had over the boys and asking how one could manipulate warring classes or nations into peace. And even if one could, OJ said, he didn't believe those kinds of alliances lasted.

I thought of the optimism of the ruling elite in Turkey in the early years of the Republic, when the new Turkishness was supposed to be the glue that bound new citizens together, and how in contemporary times it has come unstuck with the rise of conservative Islamist politics and the autocratic dictatorship of Erdoğan. Would Sherif have predicted the swamping of the engineered secular and modern identity, and the values of the Old World rushing in to take its place?

OJ understood that Sherif had an investment in the outcome of the experiment that went beyond scientific reputation, and that he was in some grand way trying to right a wrong. But OJ knew very little about Sherif's background in Turkey. And he never asked. 'It was one of those things. I had the greatest respect for him, but I felt I needed to keep him at a distance.'

Sherif's daughters had also told me how little information their father had volunteered about Turkey, how they had to coax it out of him. And I wondered about the burden of carrying traumatic experiences. Visiting Turkey, I became acutely aware of the silences that must have haunted people who lived through those early years of the Republic.

Turkish historian Leyla Neyzi has described how fear has inhibited many older Turkish people from passing on oral histories to the next generation, especially when their personal stories differ from the official version. In one interview with a ninety-four-year-old woman in Izmir, who had lived there since childhood, Neyzi noticed the silences in the woman's narrative:

> [i]t has no doubt to do with the widespread violence between Christians and Muslims and the bloodletting that occurred at the end of Turkey's 'War of Independence' that made the Turks, first victims, then perpetrators, to want to forget their suffering, and subsequently their guilt, for what they made others suffer. It is, in a nutshell, the story of modern Turkey — and remains so. Unfortunately, today's internecine violence between Turks and Kurds is a repeat performance in contemporary disguise.

Victim, perpetrator, enemy, friend. I could see how pain and guilt could drive a kind of collective amnesia, or an unofficial silence.

Yet Sertan Batur, a Turkish scholar and expert on Sherif I met in Vienna, doesn't believe that the camp studies accurately reflected the conflicts of Sherif's Turkish childhood. 'It was very idealistic,' Sertan said. 'But it wasn't realistic.' He is dark-eyed and rather serious, and when we met, he greeted me by putting a hand over his heart. We were in the wood-panelled Café Hawelka, in Vienna's Jewish quarter.

The Robbers Cave experimental design wasn't much like the Turkey Sherif grew up in, Sertan went on. 'He said he was affected by the conflicts between the Turks and the Greeks. But his research didn't capture the inequalities in power relations between the Greeks and the Turks, or the history between them.'

After all, Sherif's boys had been chosen for their homogeneity. They shared the same skin colour, ethnic background, language, culture, and religion. As such, Sherif's theory of the resolution of group conflict was of little use in Turkey, then or now, Sertan said. 'It doesn't answer problems of conflict between people of different races, ethnicity, or culture, and he did not address the issue of power imbalances between groups.' While Sertan, who works with migrant families in Vienna, admires Sherif, he said Sherif's theory 'doesn't explain big social issues. He cannot explain society for me.'

'I think he wanted to find something optimistic in the climate of the Cold War — I think he wanted to produce something politically useful so his research said something powerful and inspiring for the oppressed. He could say, "I have discovered something that will stop people hurting each other. If we focus on bigger issues, things that are important to all of us, we can overcome these problems between us." The political implications were more important than the research techniques and methods. I think he ignored his results from the early experiment and censored his own research and emphasised the optimistic rather than the realistic results.'

Perhaps Sertan was right: the Robbers Cave experiment was a Cold War bedtime story to give people hope in a time of fear — or a narrative that Sherif told himself to increase his faith in humanity and feel that he was making a valuable contribution to the future of the world. But it felt deeper than that to me.

In 1954, with the successful completion of the Robbers Cave study, the tight-knit research team dissolved. Within months, Jack White had moved to the University of Utah, in Salt Lake City, and OJ Harvey was on his way to the University of Colorado, in Boulder, via Yale. Only Bob Hood was left, and his dissertation was almost finished.

With Harvey gone, Hood became the focus of Sherif's affection. Sherif was possessive and proud of the achievements of his graduate students, but particularly OJ Harvey, who was part Choctaw; Jack White, who was Kiowa; and Bob Hood, who was Cherokee. Hood jokingly called Sherif's habit of bragging to his Ivy League colleagues about what he could do with poor boys down south as his Pygmalion complex, even though Hood's family owned a string of pharmacies in Oklahoma.

Sherif was reluctant to let Hood go. Whenever Hood tried to arrange his final oral examination, Sherif would put him off. The story goes that Sherif was so reluctant to see him leave that Hood, who had a reputation as an unconventional thinker, had to take extreme measures to gain his independence. He rang Sherif in his office one morning and said, 'I'll be on the bleachers at Owen Field at three this afternoon, and I want you to be there.' When Sherif showed up, Hood pulled out a Colt .45 and put it down on the seat between them and said, 'I want to finish my dissertation.' Sherif nodded and said, 'Okay, Bob,' before getting up and walking away. Hood graduated soon after. I laughed when I heard this story — it seems almost certainly apocryphal — but it has a disturbing edge.

With Harvey already at Yale and White making preparations to leave, the men worked quickly to write a book about the experiment. Carolyn, at home with the children during the Robbers Cave camp, had prepared what she could of the manuscript ready for the others to add the missing pieces.

But OJ was never proud of it. 'We wrote it in a hurry. And it shows,' he told me. Six weeks after the experiment was over, a photocopied report, bound in a pale blue cardboard cover and titled *Intergroup Conflict and Cooperation*, was produced in the University of Oklahoma's print room. Sherif immediately sent one hundred copies out to colleagues in psychology and sociology departments across the country, and signed a contract with Doubleday to publish it as a book entitled *Friendship and Friction Between Groups*.

Yet after Sherif sent the manuscript to Doubleday editor Josephine Lees, she wrote to him in February 1955, 'We feel that the manuscript you have presented to us is, frankly, not ready for publication in its present form.' It was clear that the chapters had been written by different authors, and at different times. 'As a result, the work does not hang together,' she noted. She rejected the photos as not being of suitable reproduction quality and pointed out that if they were to publish any photos of the boys, Sherif would have to 'obtain written permission from the parents to do so'.

The letter was like a red rag to a bull. Sherif reacted by putting his foot down, refusing to make any changes. Doubleday cancelled the contract.

The University of Oklahoma had established an Institute of

Group Relations, headed up by Sherif, in 1955, and he felt very much at home there. But by 1958, he still had not been offered tenure, despite being there almost ten years, and the university's strict anti-nepotism rules made it difficult for Carolyn to find ongoing work. In 1958 and 1959, he took leave to go to Texas with Carolyn, where she completed her PhD. It was unusual at the time for a father to be the primary parent involved in childcare, but Sherif's daughter Sue remembers it as a happy period, with Muzafer spending the summers taking his three daughters to swimming pools and the cinema, where they went to repeated showings of the Marilyn Monroe vehicle *Some Like It Hot*.

In 1960, the University of Oklahoma was considering offering Sherif the post of research professor, and the dean of the graduate college, Lloyd Swearingen, wrote to Gordon Allport: 'In your opinion has he made truly outstanding research contributions in the field? Is he truly outstanding among the men of his field in the United States?' If Allport could give them a 'frank and sincere answer', it would be held in the 'strictest confidence'.

Allport replied at length:

> He has constantly been an intense and productive scholar. Even as a student he was a tenacious eager-beaver, holding the highest standards over himself. His level of 'drive' has remained continuously high. From my point of view he deserves extra credit for tackling the really tough type of problem in social psychology, and not settling for easy and methodologically 'safe' topics.
>
> Yes, I think he would be called 'distinguished' by the majority of social psychologists in this country. His researches

> and theoretical writing have aroused much attention and are constantly cited. Professionally he 'gets around' and at meetings one can see him arguing vigorously in some corner with colleagues from a variety of institutions.
>
> If I have any hesitation, it concerns his personality. Sometimes he has seemed difficult to work with. He is somewhat intolerant of other points of view, and keeps his own pace rather than adapting to a team. But you know far better than I whether these traits that I noted in past years are any sense disqualifying. I realise that the honor you are considering is in recognition of scholarly productiveness and not of charm. I am personally rather critical of Sherif's manners, but even I would endorse the choice, if you decide to make it.

Sherif was offered and accepted the position and remained at the University of Oklahoma until 1966, when he took leave of absence so he and Carolyn could take up one-year visiting professorships at Penn State. The temporary positions turned into permanent ones, and Sherif would never work in Oklahoma again.

No one knows exactly where or when Sherif's illness started — whether it began in Turkey or if it was triggered in America — but there are hints of it along the way: his recklessness with money, his impulsiveness, his seeming imperviousness to danger, the bouts of furious energy alternating with troughs of depression. Sherif's children have little memory of him being ill when they were small, remembering a warm and loving, funny father. But Carolyn saw the roller-coaster of his highs and lows, and it must have been frightening for them both. In the 1950s and 1960s, mental illness was stigmatising, and Sherif was reluctant to seek treatment. Carolyn did her best, but it was a struggle without professional help.

Sherif's graduate students from this period remember he and Carolyn as unfailingly generous and interested in their work, always hospitable, hosting parties and get-togethers and treating them like family. Some engaged couples learned not to show surprise when Sherif gave them autographed copies of his books as wedding presents. He could be gregarious and charming, when he wasn't morose or depressed. His students saw him as indomitable and eccentric. Colleagues, however, increasingly had their reservations about his effectiveness as an educator. Sherif's former supervisor Gardner Murphy wrote to him in 1956 saying he had something to communicate that Sherif probably wouldn't like to hear:

> A considerable number of people have told me within the last few years that you have given public addresses which would be very effectual, if they were carefully planned and completed within the time allowance, and if you stuck to your text. They say you frequently leave your notes and become excited and ramble and that you frequently go on an hour or more beyond the time which has been strictly defined as closing hour.

Murphy's letter reminded me of some of the critical comments some of Sherif's colleagues had made about him around the time of the FBI investigation. Perhaps Murphy, who knew Sherif well, thought that his former student could make use of this feedback to improve his public speaking. Or did he sense that underneath this arrogance and passionate excitability there was perhaps evidence of irrationality and disorganised thinking? I wondered how Sherif, who held grudges for years against people he felt had slighted him — he carried a mildly critical review of one of his books around for twenty years and periodically urged one of his graduate students to write to the reviewer to 'set the record straight' — would have taken Murphy's letter. There's no reply in the archives and no evidence of whether it had any impact on Sherif's behaviour. But their warm correspondence continued over the years, and Sherif seems not to have taken offence. Ten years after Murphy's letter Sherif was renowned at the University of Oklahoma for evening seminars that went way beyond the allotted time. As for Carolyn, I wondered if she knew of the letter, and if she felt defensive on her husband's behalf or secretly relieved that Murphy had put his finger on something that worried her too.

In 1966, after seventeen years at the University of Oklahoma, Sherif was offered a position at Penn State, along with Carolyn, and they left Oklahoma for good. But after their move, things began unravelling.

For someone whose identity was so tied up with his work, the move from Oklahoma was a wrench for Sherif. Penn State was not the University of Oklahoma; he was no longer a big fish in a small pond. Small-group research was becoming unfashionable in the wake of the Cold War, and funding was drying up. And while Carolyn settled in and soon developed a happy and successful professional life, Muzafer's behaviour made it increasingly hard for him to make new friends or hold onto old ones. He became very jealous and aggressive if he thought a man was paying Carolyn too much attention, and one colleague described how at one conference Sherif physically attacked a man he thought was flirting with her. To Carolyn's dismay, he entertained his students with stories of eating dog food when there was nothing to eat at home. Physically, he was in poor shape. Years of heavy smoking and drinking and untreated diabetes caught up with him. In November 1968, he had a car accident under the influence of alcohol, and soon after had a stroke. It was then that the doctors looking at his brain scan asked Carolyn about earlier head trauma, and she remembered the story Sherif had told of being kicked by a camel as a boy.

In November 1969, Carolyn made some notes on some loose sheets of paper as if gathering her thoughts. She described Sherif's drinking and aggression towards her, his paranoia, and his seemingly unshakeable belief in his own 'delusions'. Despite her insistence that 'he must get well, then he would see things in perspective', it made no difference. She wrote, 'What does this add up to? I am too terrified to say. The symptoms add up to a megalomaniac ... who sees everyone plotting against him.'

Three months later, Sherif received a letter from his lawyer brother in Turkey, who wrote to say he wanted to visit, clearly hoping to make amends. But Muzafer was still bitter and didn't want to see him. Carolyn wrote down the details of a dream she had around this time — 'I had a terrible dream. Terrible because I fear it is true.' In the dream, she wrote a reply to Muzafer's brother: 'Don't let your conscience as a brother trouble you. I have been living with a crazy man — a mad man — for 25 years. The madness has only become worse until it affects every aspect of life ...'

Despite these personal pressures, Carolyn's career was flourishing. Penn State had offered her a position as part of its efforts to employ more female faculty, and even though she felt like the 'token woman' in the psychology department, she began to gain recognition for the twenty years of research and co-authorship of more than five books with Sherif. But at the same time Sherif's career languished. His bouts of depression got deeper and more frequent, and he was hospitalised several times. Stanley Milgram arranged for Sherif to join him at City University of New York, organising a visiting professorship, but at the last minute Sherif dropped out. When a former student wrote to Sherif from Turkey, asking if he was interested in a position at the Middle East Technical University in Ankara, Sherif never replied, although he kept the letter until the end of his life.

He had been cut adrift from Turkey. He had no passport, so no official proof that he had ever been a Turkish citizen. Sherif's daughter Ann said he was often embarrassed to speak the language because it had changed so much that his Turkish sounded antiquated. I imagined how painful it must have been to be marked out a relic of the past, when he'd begun as a symbol of Turkey's future. But he never stopped thinking about Turkey. He loved travelling, and on family trips in Northern New Mexico or Texas or Arizona, he often stopped at some scenic or wild spot to exclaim to his daughters, 'Oh, babies, this reminds me of Turkey!' As he aged, he became increasingly nostalgic for his country of origin, but remained paranoid about whether he would be safe in going back. He didn't know what kind of welcome, if any, he would get. He was bitter that his older brother had had him legally disowned from the family, that his colleagues at the university in Ankara had not come to his aid when he was jailed. He was never able to shake off the suspicion and paranoia, and perhaps partly understandably: in Turkey he had been under surveillance and informed on, and in America he had been investigated by the FBI.

He worried especially that the Turkish government might still be out to punish him. One of his colleagues from Ankara University, folklorist Pertev Boratav, who had been jailed for undermining nationalism and 'promoting leftism' in his classes at Ankara University, had moved to France upon his release, in 1952. In 1974, the Turkish Ministry of Culture inaugurated an International Folklore Conference and accepted Boratav's submission to present, but at the last minute his paper was removed from the program.

Sherif didn't renounce his Turkish citizenship, but he didn't take up US citizenship, either. This meant he couldn't vote, but it also meant he couldn't travel to international conferences because he had no passport. Although his daughter Sue said he quipped that he was a 'citizen of the world', without citizenship or a passport of any country, his status meant he remained an outsider.

I wondered at what point Sherif realised or admitted to himself that he would never go back to Turkey. Perhaps it was towards the end of his life, when his daughters were able to coax stories about his childhood from him. But for a long time he felt a bitter sense of betrayal for being unexpectedly exiled from his homeland. And talking about Turkey must have been a painful reminder of what he had lost.

The world was slow to catch on to what Muzafer Sherif already knew: he was a giant in social psychology. And until people acknowledged the fact, he chafed and complained about the slowness of his well-deserved recognition. But underneath this bluster and bravado, OJ Harvey told me, Sherif yearned for acceptance.

'He always felt like an outsider.' A horse whinnied from the paddock next door to OJ's house and OJ cocked his ear, listening for a moment, before continuing. 'When he came over here to America in 1945, he got a ride on a military plane. He still told the story of that plane ride decades later, in vivid detail, like it had just happened.' OJ shook his head. 'Some of the guys on the plane were military brass and they asked him, "What do you do?" And Muzafer told them, "I'm a social psychologist." And they couldn't believe it — a Turk being a psychologist!' OJ's voice went up and he waved his arms, imitating Sherif's outrage. Sherif never got over it, OJ said. A decade later, Sherif was still bitter, recounting the story of their incredulity and prejudice and his humiliation. He had this craving, OJ said. Sherif felt that nobody understood him, or took his work seriously enough. He wanted recognition of his genius.

'He was terribly bright. But despite being so bright and well educated, he never felt equal. Isn't that amazing?' OJ said this with a kind of wonder. 'He viewed himself as top of the heap — oh, no question. But he felt he never got the accolades he deserved. He was his own worst enemy that way. He thought it was prejudice that meant he didn't get recognition he deserved. I mentioned to him a couple of times — you know, if he hadn't been so insistent then maybe people would have been more generous in their recognition of his ability? But that didn't change him.' OJ laughed. 'Maybe it was true that people begrudged him awards. He could be very competitive, he could be very difficult, and so people had very mixed feelings about him.'

In the late 1960s, Sherif was presented with both the prestigious Kurt Lewin Memorial Award and the APA Distinguished Scientific Contribution Award. But he seemed incapable of enjoying the honours and made his displeasure known. Two of his former graduate students wrote that he complained the award citation was sloppily written and incorrect, he should have gotten the award sooner, and he wondered why so many social psychologists he regarded as inferior had been awarded it before him.

By now Sherif's intellectual influence was waning. In 1948, he and Carolyn had published *An Outline of Social Psychology*, a well-received textbook. But a subsequent revision in 1956 was less successful, and the final revision in 1969 was even less so, seemingly outdated as it reflected their own particular view of social psychology rather than the rest of the ever-expanding field. Bob Hood used their textbook when he moved from the University of Oklahoma and taught at a women's college in Pittsburgh. One of his students that I spoke to told me, 'I went to graduate school in social psychology having no idea that there were any other social psychologists in the world other than the Sherifs. But at graduate school I discovered all these new theories and ways of looking at the world. And I called Bob and I was furious.' Several of Sherif's former graduate students agree that the Sherifs' book looked 'selective and unrepresentative', but suggest that before the 1960s this was typical of social scientists, who used their texts to promote their own theories and cite their own research.

By 1967, there were only three rather tatty-looking copies of *The Robbers Cave Experiment* left for sale in the University of Oklahoma's bookstore, where it had sold 5,000 copies over the previous thirteen years. But when Bob Hood approached them about reprinting it, they refused on the grounds that Sherif had left the university — and also, Hood suspected, processing orders and distribution was a 'headache'. The bookshop declared it out of print. The Sherifs had kept the story of the experiment alive in their books and articles published during this period.

Between 1970 and 1977, Carolyn, perhaps in an attempt to lift Muzafer's flagging spirits as much as to maintain the profile of their research, tried to get the book published. They sent out a proposal to more than twenty publishers, pointing out they received a steady stream of requests for copies of the now out-of-print monolith and that 'most introductory books in psychology, sociology and social psychology include the experiment'. But they had no success, even though by then the Robbers Cave study was regarded as something of a classic.

In 1977, the CIA released thousands of documents, after a freedom of information lawsuit, about its funding of research into mind control and interrogation techniques that could be used against enemies. Carolyn was horrified to find that Muzafer had unwittingly accepted funding from the CIA for small-group research he conducted while she had been completing her PhD a decade earlier. Sherif had conducted a covert observational study on groups of adolescent gangs. It was part of a program of top-secret experiments called MKUltra. But while Sherif was studying urban gang members, the CIA applied the same research to techniques for renegade members of the KGB: 'Now, getting a juvenile delinquent defector was motivationally not all that much different from getting a Soviet one.'

The Sherifs weren't the only social scientists duped by the CIA, but the news must have been distressing, given their idealism and political views. At a professional forum and in the pages of the *APA Monitor*, Carolyn reiterated that she and Muzafer 'did not share the Cold War consensus' and had no idea that the funds had come from the CIA.

It was likely they shared the same sense of incredulity and outrage expressed by sociologist Jay Schulman, who wrote:

> … it had to do with my own naivete. Even though my politics were socialist, I had no understanding at that time of how the real world operated … In 1957, I was myself a quasi-Marxist and if I had known that the study was sponsored by the CIA, there is really, obviously, no way that I would have been associated with that study or that work … My view is that social scientists have a deep personal responsibility for questioning the sources of funding, and the fact that I didn't do it at the time was simply, in my judgment, indication of my own naivete and political innocence in spite of my ideological bent.

However, Carolyn's career continued to blossom. At Penn State she was quickly promoted from temporary faculty to tenure track and from associate to professor within four years. She wrote in an autobiographical essay that, after years of teaching and research and writing with Muzafer, first at Princeton, then at Oklahoma, finally 'academia made room for me'. She had had what one friend called 'an epiphany' about her own experiences as a woman psychologist, and gender discrimination became her intellectual passion:

> To me, the atmosphere created by the women's movement was like breathing fresh air after years of gasping for breath. If anyone believes that I credit it too much for changes in my own life, I have only this reply: I know I did not become a significantly better social psychologist between 1969–1972, but I surely was treated as a better social psychologist.

But even before he retired in 1972, Muzafer was often depressed, and friends recall how Carolyn tried to keep him engaged, inviting students over for meals and parties. Most times, Sherif stayed in his room instead of coming out to join them. He had by now been diagnosed with bipolar disorder, and he was in hospital again in 1982 when Carolyn was admitted with a sudden illness. And he was still in hospital when she died of cancer, aged sixty, just a few months later.

Carolyn had been his lover and wife, his co-author and intellectual partner, his voice of reason, his port in the storm, his carer. He was devoted to her. During one of the group experiments, he wrote to tell her how he felt after she had called him one night: 'In the midst of all this, your voice … adds so much beauty to … the terribly complex circumstances of my life and … keeps me standing on my feet.'

Carolyn had shared his idealism, his belief that social

psychology was a calling, his great faith that they could do something important for humanity, and a steadfast belief in his brilliance. They talked social psychology and ideas at the dinner table, at the beach, on family camping holidays. In her own way, Carolyn was as driven as he had been.

Without her, he was devastated. He was in and out of hospital over the next three years, and attempted suicide at least once during this time. It seemed impossible that he would recover.

Just before his eightieth birthday in 1986, Sherif and three of his graduate students from Penn State took a road trip to Robbers Cave. He was on a lithium regime by then and, according to his family, a changed person. The new Muzafer was mild-mannered and moderate, rather than extravagant and opinionated. Sherif talked social psychology most of the way, but at the park, he was quiet for a while. Then he took the students on a guided tour, pointing out the cabin where after the experiment he and Carolyn used to come and stay on vacation with their daughters, before moving deeper into the park to find the mess hall, the baseball pitch, and, of course, the cave.

At the park's entrance, by the pile of logs carved with white letters spelling out 'Robbers Cave State Park', Sherif sat and posed for a photo. He had a copy made and later sent it to OJ with a note saying how great it was to revisit the place after all those years — the note where he thanked him again for making the experiment happen.

It took me a moment to recognise the old man in the photo when OJ showed it to me. His cheeks were hollow; his grey hair was long and straggly. But his ravaged face looked peaceful.

A year later, in 1987, OJ began making arrangements for the book they had written about Robbers Cave to be published. Perhaps because Sherif had so recently written enclosing his photo at the state park, or maybe simply because OJ knew how pleased it would make him, he wrote to Sherif to let him know. Sherif's response was jubilant:

> Dear O.J. — old friend!
>
> Warm thanks for your letter of April 13. I do appreciate the close feelings conveyed in it. It is high time that at last a university press decided to publish the Robbers Cave … I am also very happy that the introduction is to be written by Don Campbell and not by any lesser person in theory and methodology … With heartfelt appreciation for all your activities in Norman and since then … Sincerely, Muzafer.

I don't know if Sherif saw the book before he died the following year — whether he held the slim volume in his hands and turned it over to see the impressionistic drawing of the boys engaged in tug of war on the front cover. It hardly mattered. It would have been enough for him to know that finally, over thirty years since the research at Robbers Cave, their work stood alone. I could imagine the frisson of pleasure he would have felt reading the words of Don Campbell, a psychologist he so admired, emblazoned on the book's cover: 'There have been no subsequent studies of anywhere near the magnitude of the Robbers Cave experiment …'

The next time I went to Robbers Cave, I climbed the track up to the top and sat on the sandstone rock to look out across the valley. It was mid-morning and the colours were still deep. I closed my eyes. Up here on top of the high ledge, with a breeze ruffling the trees and my thoughts full of outlaws, I could be on Mount Bozdağ, with its own robbers caves, its stories of bandits hiding in the hills. I thought of the moment in the experiment where Sherif knew the thrill of success — when the formerly pious and timid Eagles threw themselves in fury at the enemy group, the Rattlers. What prompted it was not competition, as Sherif argued, but the theft of the boys' knives. It was an act of banditry.

I remembered the story of the new mother on Mount Bozdağ the year that Sherif was born, and her terror hearing that her baby would be stolen from her by the brigand Charkirge and his outlaw gang. And how, down in the main street of Bozdağ today, a statue of a bandit Poslu Mestan Efe now stands, he and those like him now immortalised as heroes for the guerilla war they conducted against the invading Greek army. Muzafer Sherif's childhood was marked by these violent reversals. In the absence of a Turkish army, the lawbreakers and bandits in Ottoman times became heroes of the new Republic, resistance warriors who marched with Atatürk at the head of parades, receiving military rank and pensions for their services. I couldn't help thinking that perhaps Sherif hoped for his own kind of reversal, where instead of being shunned and marginalised in his homeland, he could be welcomed and feted on his return.

I had found some notes Sherif had made for a talk at Princeton around 1945, soon after he arrived back in America:

> Perhaps I am the first Turk you ever saw in your lives. But you have undoubtedly heard about the Turks … you surely must have listened to the stories about us — 'the Terrible Turk' — with a dagger in his hand killing people whenever he can, starving the poor Armenians etc. etc. … That is not the Turk I know.

In a series of bullet points, Sherif jotted down evidence of the modernisation of Turkey — education, dress, transport, and literacy, his attempts to counter his audience's prejudice with facts, as a way of changing their minds. The Turk he knew was civilised, educated, forward-thinking, and passionately patriotic about his country.

I had gone back and forth in my thoughts on Sherif's motivations during my research, discarding one hypothesis then another. Sherif's camp studies, including the final and successful Robbers Cave experiment, were a mirror of his past; a celebration of his ideals; a tribute to the power of collectivism; a testament to Marxism; a triumph of social engineering. Perhaps it was all these things, a historical canvas onto which I could project any number of hypotheses or explanations.

But knowing what I knew now about Sherif's persecution and exile, the experiment's exploration of themes of friendship and betrayal, identity and belonging, and the call to let go of rivalries and bring enemies into the fold, made it seem like a love letter to the young Republic, the early years of the new Turkey.

Perhaps it was in the Ottoman period, on Mount Bozdağ, that Sherif experienced his earliest and most powerful notion of reconciliation. A week after the picnic the young mother had held in his honour, the bandit Charkirge returned the invitation. On the plateau, almost at the top of the mount, the Levantine families arrived to find a lavish spread, with roasting lamb, pilaf, yoghurt, and sweet kadayif. Afterwards, all the guests gathered wild blackberries together for the families to take home and make jam, and instead of her earlier anxiety, the young mother felt 'comfortable' with the bandits and their families. The crisis was averted; the baby boy was safe. Her fear evaporated. 'Now we were friends,' she wrote.

Down on the path I heard a boy laugh. Someone called out 'Echo echo echo,' and the call travelled out from the ledge and back again, boys' voices bouncing and calling from the present and the past.

What was the Robbers Cave experiment about? What were they trying to prove? All of the boys I have met and interviewed have asked me the same question. And I have struggled with an answer. It was about groups and fighting, and then groups making peace, I said, but it sounded weak. Is there any answer I can give that doesn't sound like a truism? And they sense it's not as simple as that, either — some wonder if they have been part of someone else's ethical drama. The experiment might have been a metaphor for nations or countries or ideologies in conflict, but since I met some of the boys and they realised that they were unknowing participants in an experiment, they have begun a struggle to understand its moral implications and what effect it has had in their lives. Some, like Doug, have made a kind of peace with it and moved on. For others, that journey is just beginning. The now-adult boys I've spoken to are still affected in their own ways by being in Sherif's experiment. As I was drafting this chapter, one of them emailed me, wanting to know when the book was coming out. 'Can't wait to share with family and friends so they will understand why I am a bit odd,' he wrote, adding a smiley-face emoji to show he was joking. Or was he?

Reckless and cautious, egalitarian and elitist, last generation of an empire and first of the new Republic, adored husband, often infuriating colleague, loving father, driven man. I have hoped to shape a figure from the slippery clay of anecdotes and actions, from gossip and snatches of history from a forgotten era. Running through it all, a thrumming thread of energy, was Sherif's unshakeable belief in a theory of power of tribal loyalty, in-groups and out-groups, that reflected his own experience as an outsider looking for a place to belong.

I had read Orhan Pamuk's memoir *Istanbul* in preparation for my trip to Turkey. In it, he evoked the Istanbul of his childhood, haunted by 'hazun': a deep sense of loss and melancholy; the hollow left by a disappearing world, and an intense nostalgia for the glory days of the Empire. Istanbul, once the centre of the Ottoman Empire, had been abandoned as the capital, after the seat of power moved to Ankara.

In Istanbul, I had walked around in circles, up and down hills in the city's Beyoğlu district, looking for Pamuk's Museum of Innocence, a building that Pamuk planned along with the novel of the same name.

When I finally found it, I wondered how on earth I'd missed it. A tall traditional wooden house with an overhanging bay window, it's painted a bright modern red and sits dramatically on the corner of a steep street. Pamuk bought the building in the 1990s, when the area was rundown and the building was dilapidated. Now the neighbourhood has sprouted fashionable coffee houses and galleries. But his novel was finished and published well before the renovation of the building was complete, and the museum did not finally open until 2012.

Inside the Museum of Innocence, Pamuk has gathered together everyday objects and organised them to tell the story of the characters of his novel. A salt shaker, playing cards, lottery tickets, soap, a cinch-waisted 1950s dress covered with roses, movie posters, identity papers, newspaper clippings displayed in glass cases — all document the story of a failed romance and provide a vivid picture of life in post-1950s Istanbul. I smiled to myself, realising that in coming to Turkey I had naïvely hoped to find something similar to tell me as much about the everyday life, the emotions, and the experiences of a young Muzafer Sherif. As if such a thing could have existed.

In the lobby, what at first seemed to be a wall of cuneiform script turned out, when I stepped closer, to be an entire wall covered with over 4,000 cigarette butts, some smudged with lipstick, all skewered with pins and annotated with the date and place each was smoked by the narrator's beloved.

I spent the afternoon in the museum, wondering at its obsessiveness and its charm and its moving depiction of the lives of the characters. Pamuk had created an entire world, right down to the loft bedroom where Kemal, the narrator, told his story. It was a self-contained world, referring to the particulars of a specific time and place. I wondered at its almost complete absence of any reference to Turkish politics.

Perhaps it was that I was still looking for connections, but in a strange way, the narrow focus, the obsession with detail, the careful avoidance of any direct political commentary reminded me of Sherif's work at Robbers Cave, as if the experiment was a novel, the earlier experiments drafts of the later, completed opus. Perhaps Sherif's experiment wasn't just a metaphor about Cold War politics or the idealism of the Kemalist years. It was more personal than that; it was more like the kind of story that people tell themselves to make sense of wars and violence, a narrative with heroes and villains. Throughout his life, Sherif was emotionally conflicted, struggling with competing and sometimes violent feelings of love and hate, trust and suspicion, sanity and madness. At Robbers Cave, he had created a perfect moment, or recaptured an old one: a world where wounds were healed and what was lost was restored, a place where all was whole and complete, while in the world outside, things were falling apart.

When the young woman called to me from the bottom of the stairs that the museum would be closing in fifteen minutes, I didn't want to leave, to go outside again. It felt safe here, surrounded by these objects, caught in the embrace of this unfolding story.

# A Note on Sources

On a high shelf in a vast back room at the Centre for the History of Psychology at the University of Akron is a wooden trunk. A label dangles from it addressed to Muzafer Sherif, 728 Chatacqua Street, Oklahoma. The wood looks battered, and it's likely the same trunk that Sherif first brought with him from Turkey, containing a few changes of clothes, some photographs, and his precious psychology books. Later, he used it to store different mementoes, and the props and paperwork for his most famous experiment. Sherif's science is like that trunk — its beginnings far from rural Oklahoma, tangled up in the life of Muzafer Sherif, the dissolution of an empire, and the traumatic birth of a new nation.

In writing this book I wanted to recreate the backstage world of the Robbers Cave experiment, to explain where it came from, how it started, and why when it ended, Muzafer Sherif seemed to fade away. It's been a slow and often frustrating research process. While I have had access to detailed scientific records describing how the boys were recruited, and daily descriptions of particular behaviour Sherif and his team were looking for, the men's observer notes often make for dull reading. The flesh and blood of this story — the lived experience of the boys themselves — appears only in brief flashes.

When Dr David Baker at the Archives of the History of American Psychology at Akron first told me excitedly that the 'gem' of their collection was the newly acquired material from the Robbers Cave experiment, I told myself the trail was too cold. The boys, and who knew how many of them were still alive, would be impossible to find. Sherif had died, hence his family's donation of the material to the archives. And I, in the middle of researching a book about Stanley Milgram's research, didn't have the time or the energy to pursue it.

David also told me how one of Sherif's assistants at the Robbers Cave, OJ Harvey, had recently visited the archives to help with the cataloguing of the collection. It had been an emotional visit for Harvey. Sometimes he wept, sifting through the papers and photographs and letters. At the university he gave a couple of guest lectures about the experiment, which were received enthusiastically by the students. Perhaps it was the students' interest in his behind-the-scenes reminiscences, or his awareness that he was the last researcher from Sherif's famous study still alive, or that with Sherif gone he felt free to tell his version. But OJ was keen to tell the story, David told me. And because David and his staff have been so good to me over the years, I felt bad turning such an opportunity down. I travelled to Boulder in May 2010 with my recording equipment and spent three days interviewing OJ Harvey. By the time I left, I was hooked.

The French philosopher and sociologist of science Bruno Latour refers to the 'Janus face' of science. On one side you have the public view, the face that scientists and researchers want us to see, the official frontstage presentation of facts by objective-technologists-in-white-coats. It's the PR face. Then you have the other face, the view backstage. It is messier, sometimes ugly, but always more interesting. From here you can see how and why science — and in this case, social psychological science — is made. It's a world of human beings, not impersonal experts in lab coats, but an array of individual characters with personal histories and emotions that rarely make their way into public view.

In public accounts of the experiment, the individual boys who took part in Sherif's research get lost in discussions of Sherif's theory of group conflict and applications of the research. They cease to be children and become stand-ins for countries, ethnicities, ideologies, humankind. But they were individual boys who brought their own backgrounds and expectations, family histories and hopes, to that camp in the summer of 1954.

The boys in Sherif's camp experiments are lost in many ways. While there are hundreds of pages of observations about them in the archives, the commentary is often about the group as a whole, rather than individuals, or about leaders rather than followers. When a single boy is mentioned, his name has been redacted. When Sherif or one of his assistants describe a disagreement between boys, they don't often include information about what prompted it or how the fight developed. I didn't want to write a book in which the boys were an undifferentiated group of faceless individuals with a hive mind.

From the start of my research, when I somewhat reluctantly walked into OJ Harvey's home in Boulder, Colorado, in May 2010, to the time I finished in December 2017, I knitted together the delicate tatters of the narrative from the accounts of research staff, letters and diaries, the recollections of some adult boys, and the observations of people who knew Sherif or are deeply familiar with his work. But it was not an easy or straightforward process. Although Sherif wrote prodigiously, it is no exaggeration to say that social psychology was his life. He is an elusive character, and while his personality is alive in his letters and writings, it's rare for him to mention his thoughts and feelings.

I followed different leads. I made contact with Muzafer Sherif's daughters and made tentative plans to meet up with his eldest daughter, Sue Sherif, in Las Vegas, where she was attending a conference. But it didn't work out and instead we talked on Skype, she from her home in Alaska, where her father was living with her when he died in 1988. Later still I called his daughter Joan, who shared her memories of her parents, and his daughter Ann sent along a photo that captured their spirit.

I knew so little about Turkey — how did I know so little about Turkey? — that when I started to read about its history it was like trying to put the pieces of a jigsaw together without the picture on the lid of the box. I read indiscriminately, following first one strand, then another. I got lost in the Lausanne Treaty, the cult of Atatürk, the life of Ottoman sultans, a whole book on the outlawing of the fez. I read the diaries of American missionaries and diplomats, peppered with dismissive and often supercilious analyses of the shortcomings of the Turkish people. I read memoirs written by Levant ladies yearning for the lost days of Smyrna, their servants, and leisure and wealth, written from their exile in grimy brick flats in London.

But nowhere did I find a description of what life was like,

the lived experience of a boy like Muzafer Sherif, growing up in Ödemiş. I got close: in the memoirs of his headmaster and the letters of some of his teachers, I got a glimpse of the young men like him. But I could not find specific mention of the dark-eyed and intense Muzafer Sherif.

In Vienna I met with expert on Sherif's work, Sertan Batur, who hid his surprise at my ignorance and was patient and kind. I interviewed former students of Sherif's and the men he worked with, consulted the archives at a range of universities and foundations. And of course I went over and over Sherif's own records, the thousands of items related to the camp experiments. But I got despondent. I reminded myself that research was a haphazard and circuitous process and you never knew where you'd end up. But I also knew that truckloads of facts don't tell a story.

The challenge in writing this book has been how to fill the gaps in the narrative while staying true to the facts. This was a particular challenge in writing about the camp experiments. I have used Sherif's audio recordings and photographs as a framework and anchor points for the events described in these pages, as well as the notes of the participant observers, whose job it was to record their observations of the boys in each group. The observers' notes have largely been redacted, although the observers paid more attention to what they called 'high status' boys in each group, and to noting the interactions between leaders and 'lieutenants', so I have been able at times to infer which boys the observers are referring to. In other places it is impossible to tell which specific boys are being described.

In forming a narrative thread to connect events in some places, I have recreated events through a combination of the official record and interview material. In places I have added dialogue or gestures, or speculated on the thoughts and feelings of the people involved, or provided elements of a boy's backstory to enable the reader to follow the story of individual boys in the experiment. When I have added dialogue or the thoughts and feelings of boys, I have tried as much as possible to base this closely on interviews and archival material, as well as conversations with the now-adult boys. At other times I have inferred the boys' emotions from observations or written comments from the adults.

Then there is the question of memory. I was able to track down and talk to some of the boys, but often their recall of events was minimal, hazy, or incomplete. In 1954, these men were bright, often boisterous, and sharp-eyed children. But sixty years on, recalling their childhood selves and pinpointing their specific thoughts and feelings at the time often proved impossible. Complicating this was the fact that they were often grappling with new information about the experiment that challenged their previously held points of view. Having said that, I found the adult boys' recall of events proved surprisingly consistent with those provided through the official record.

As for the impact of the experiment and its effect on the now-adult boys, I have no way of knowing whether they had rewritten the narrative over the years to reconcile themselves to the experience, or if their lack of recall was a defence against remembering unhappy events. After all, these now-grown boys were sixty years ago faced with dilemmas in an experiment that was not of their choosing.

I make no claim to a representative sample. I spoke to those boys who I could find and who were willing to speak to me. Some asked me to use a different name and I have done this. All of those who participated did so because, like me, they were fascinated by the thought of this kind of experiment and of their role in history.

In trying to answer the questions I had, my research took me to Ödemiş and Istanbul; to Vienna; to upstate New York and rural Oklahoma — from contemporary Turkey to the Ottoman Empire, the Turkish Republic and Mustafa Kemal and back again. I am not a historian, and I have begun with a superficial knowledge of Turkish history. While I have consulted experts in this field where possible, any errors of interpretation or fact are mine alone.

In writing about Sherif, I have abandoned the idea of him as the faceless scientist backstage and inserted him as central to the story of the experiment.

The backstage view of social psychological research, I believe, helps us to understand not just how we come to accept such research as fact or the role of the experimenter's own life in shaping it. It also goes some way towards answering the question of what is the cost to the subjects in psychological research? Can people experimented on as children ever emerge unscathed?

Getting to know individuals who were subjects in Stanley Milgram's obedience experiments taught me that subjects in psychological research are not passive beings who unquestioningly accept a scientist's instructions — those subjects are a social psychologist's fantasy. My hunch was that the boys, now men, from Sherif's studies could offer insights into the experimenters and the research that the adults would have missed. But giving equal weight to the stories of the scientists and the subjects is more than an attempt to redress the power imbalance in accounts of Sherif's research. Social psychologists are storytellers who help explain human nature to us; but whether the tales they tell are sensational, alarming, or comforting confirmation of our deeply held beliefs, we should question their conclusions. Behind the mask of science is the art of the narrative, and accounts of social psychological research are driven as much by imaginative impulse and rhetoric as they are by logic and rationality. Using the lens of history to explore the narratives of the researchers and the researched can illuminate the gap between the ideal of science and the reality of its execution.

# Chapter Notes

**Prologue**

Description of Eagles' return to cabin, confrontation at Rattlers' cabin over missing knives, breaking up of fight and dialogue based on Sherif's notes from Muzafer and Carolyn Wood Sherif Papers, Archives of the History of American Psychology, The Drs Nicholas and Dorothy Cummings Center for the History of Psychology, The University of Akron (MCWS Papers), and from Sherif et al.'s book about Robbers Cave, 1988; men's reaction and pulling boys apart, Will's thoughts, family background from interview with OJ Harvey, May 2010; Sherif's description of 'wicked, vicious' youngsters taken from Sherif, M & Sherif, CW, 1969, p. 254.

**1: Tangled Beginnings**

Description of trip from Oklahoma, arrangements with buses, and set-up of experiment from Sherif et al., 1988 and interview with OJ Harvey, May 2010; manipulation and trickery in Milgram's experiment from Nicholson, 2011; Sherif converting Soloman Asch to social psychology from Granberg and Sarup, 1992; Sherif's *Washington Post* article, 1969; filmmaker approaching Sherif, MCWS papers; Lepore's comment about social psychological research from Lepore, 2011; description of Happy Valley camp and boys from Rohrer and Sherif, 1951, and quote 'lined up on opposite sides of the mess hall', p. 418.

**2: In the Wild**

William Golding's experiment from Carey, 2009; Mary Northway quote from Wall, 2008, p. 79; history of summer camps from Van Slyck, 2006; involvement of psychologists in summer camps and 'quickly set to work' from Doty, 1960, p.145; account of therapeutic camps from Gass, et al., 2012; Nazi summer camps from Weeks, 2015 and Hiltzik, 2014; Yale's Attitude Change Program from Simpson, 1994; 'libido' quote from letter S Flowerman to M Sherif, 9 Dec 1948; 'one whit to their elimination' quote from letter M Sherif to S Flowerman, 15 Dec 1948; Lippit's invitation to Sherif, MCWS Papers; 'happiness of their campers' quote from Herald Tribune Fresh Air Fund camp counsellors training manual, 1948, from MCWS Papers; letters to ministers, letters to parents for 1949 and 1953 studies, from MCWS Papers; description of 1953 camp for parents from 'Camp Information Bulletin', Rockefeller Foundation Papers, 1.2, Series 200S, Box 590, Folder 5051.

**3: Lost and Found**

Descriptions of Little Albert experiment, John Watson, and Rosalie Rayner from Chamberlin, 2012, Harris, 1979, and Powell et al., 2014; finding Douglas Merritte from Beck, Levinson, and Irons, 2009; conclusion that Watson knew the baby was neurologically impaired and hid the knowledge from Fridlund, Beck, Goldie, and Irons, 2012; new 'myth' in psychology texts from Powell et al., 2014; '… since Little Albert was not a healthy child' quote from Levy, 2013; conclusion that Albert was a baby called Albert Barger from Powell et al.; Sherif's footnotes about the abandoned 1953 study from Cherry, 1995; history of Schenectady and The Plot from Blackwelder, 2014; Doug Griset background and quotes from interviews and correspondence with Griset, 2011–2017.

**4: The Watchers**

Description of campsite from photos in MCWS Papers; Sussman getting desperate from letter M Sussman to M Sherif, 22 May 1953; feeling the full weight of responsibility from letter M Sussman to M Sherif, 9 July 1953; biographical detail about Sussman at Yale and as 'hardworking' from Sussman, 2001; Sussman's interest in working with Sherif from letter M Sussman to M Sherif, 21 January 1953; Sherif fired pretty nurse who would be a distraction from letter C Sherif to M Sherif, n.d. from Box 3543.1 Series 2, Folder 1; 'You must be able to imagine …' quote from letter C Sherif to M Sherif, 7 May 1953; 'Herculean' quote from letter M Sherif to M Sussman, 9 July 1953; details on how subjects were selected from 'Sample Selection' report by M Sussman, 24 May 1954; summary of jobs done by workmen from letter M Sussman to M Sherif, 26 May 1953; Kelman described as research consultant from 'Staff Policy July 1953'; 'burden of responsibility' quote from letter M Sherif to C Hovland, 23 March 1953; Rockefeller Foundation grant details from 'The Rockefeller Foundation Annual Report 1952'; forest fire to bring groups together from MCWS Papers; description of archery and dialogue reported in observer notes in MCWS Papers; Carper's experience with 'rat psychology' from interviews with OJ Harvey and Professor Howie Becker, 2015; lunch menu from camp menu notes, MCWS Papers; Harold's contentment at being alone, close observation of adults, persistence in asking questions, and suspicions about microphone in the rafters from observer notes; description of first day of camp, developing friendships, archery contests, activities in mess hall, boys playing tunes, and dialogue from observer notes, interviews, and personal correspondence with Tony Gianelli, 2014, Doug Griset, 2011–2017, Walt Burkhard, 2011–2017, and OJ Harvey, May 2010; instructions for observers prepared by Herbert Kelman from 'General Orientation and Coordination of Research Plans for Participating Staff

Members', 23 July 1953, MCWS Papers; announcement of separation of groups from 'Plan for Saturday July 25, Stage 2 Day 1 (whole day 3)', MCWS Papers; change of plan on day two to compare boys in games from observer notes; separation of boys into two groups and boys' reactions to separation from photos from MCWS Papers, interview with OJ Harvey, May 2010, and observer notes.

**5: Initiation**

Harold McDonough's observations of men and not surprised at separation from observer notes, MCWS Papers; 'pain of separation' quote from Sherif's observation notes; Mickey's running away, the search for him, and 'Please let me go home' dialogue from observer notes; description of dayhikes from observer notes, physical description of setting and body language added; Sherif's 'feeling of belongingness' and 'personal identity' quotes from 'A Preliminary Experimental Study of Intergroup Behavior', MCWS Papers; boys' curiosity about other group and separation, pestering cook, and being told the other boys 'were not to be interrupted', from observer notes; dialogue between boys about spending prize money on rubbers, women, generated from reported conversation in observer notes; boys' dialogue regarding having offended Ness added and based on Carper's observation that boys were looking for ways they might have angered him; discussion of flag emblem, decision regarding lamp from observer notes, detail about setting added; boys' curiosity about separation, conversations with cook from observer notes; boys' attempt to see their friends from observer notes and interview with OJ Harvey, May 2010, reactions of three boys to being turned away added; Sussman fixing clocks at Union College from Sussman, 2001; Sussman and church service from observer notes; conversation between Jack White and boy concerned about swearing, and OJ's initiation, from observer notes, interviews with Brian Kendall, May–Sept 2014; descriptions of three-night hike and conversations with boys from observer notes; description of White and Sherif and boys' activities back at base camp from observer notes; arrival of flags, boys' excitement, flag design and staff involvement, and provision of t-shirts and caps descriptions from observer notes and interview with OJ Harvey, May 2010; depantsing definition, Sacandaga Reservoir, strip poker, and playing on bomber plane from observer notes; argument over cooking, dialogue added based on observer notes; boy hiding and crying in bushes from observer notes; White's mental comparison of potential leaders added, his identification of Doug as potential leader based on observer notes; Lake George camping trip, Carper taking charge, group meeting, and dialogue from observer notes, interview with Walt Burkhard, 6 June 2011; hotdog dinner and phone call to Sherif from observer notes, interview with OJ Harvey, May 2010; ESP conversation and talk of mothers from observer notes; Sherif's visit to camp from observer notes, interview with OJ Harvey, May 2010; Doug waking counsellors from observer notes; conversation between Harold and Carper from observer notes; anxious conversation between research team from Sherif's notebook, interview with OJ Harvey, May 2010.

**6: Showdown**

Description of breakfast scene on morning of announcement added; Ness' announcement and dialogue from audiotape A94, tape 13, MCWS Papers; 'only the winning prize' quote from Doug Griset interviews, July 2012; Sherif's choice of knives as reward and boys' dreams of them from letter M Sherif to C Hovland, n.d., MCWS Papers; sagging spirits of Carper's group, observation of baseball game from observer notes, dialogue between boys during game added based on interviews with Tony Gianelli, February–July 2014, and interview with OJ Harvey, May 2010; threats and 'don't pay attention' dialogue from observer notes; boys' reactions when Doug hit and boys' dialogue from observer notes; description of day two of tournament, including boys' excitement over 'mystery parcel', discussion of Cobras or Pythons as group name, Carper's 'tussle' with boys from observer notes; Sherif feeling stressed, sleeplessness, and drinking from interview with OJ Harvey, May 2010; Sussman as conscientious objector from Sussman, 2001; 'thick skin' from interview with OJ Harvey, May 2010; 'for goodness sake' from letter C Sherif to M Sherif, 29 July 1953; Sherif's drafted reply 'whole universe ... whale of a project' from letter M Sherif to C Sherif, n.d., MCWS Papers; Sherif prowling camp from interview with OJ Harvey, May 2010; burning fly from observer notes; frustration exercise and staff cutting flagpole rope from interview with OJ Harvey, May 2010; conversation between boys, Ness, and Sherif reported in observer notes; 'This type of loyalty' quote from Baker, 2007, p. 111; 'cultural pattern of sportsmanship' from Kelman and Carper, 1953; flagpole rope dialogue and swearing on Bible and flag based on audiotape, with addition of boys' gestures and reactions inferred from audiotape A93, tape 14, MCWS Papers, observer notes, interview with OJ Harvey, May 2010; Sherif adding events to tournament and boys' suspicions from observer notes; conversation between Sherif, Kelman, and Doug on leaving camp from observer notes and Sherif's notebook, MCWS Papers; tournament 'rigged' and boys' attitude to Ness from observer notes; smearing table and laundry mix-up from observer notes; evidence of bullying in both groups from observer notes; difference in staff reactions to archery from observer notes; Sussman's description of fraternisation, friendliness between boys in football game from observer notes; Carper spreading the 'rumor', confrontation between Sussman and Carper from Sussman's observer notes; Carper's later depiction of the experiment as a 'joke' from interview with Prof. Howard Becker, April 2015; Carper family background from interviews and emails with Hilda Carper, April 2015, Prof. Dirk Eitzen, April 2015, Prof. William Firestone, May 2015, Joe Springer (curator, Mennonite Historical

Library), April 2015; 'fellow at Yale' quote from interview with Prof. Michael Lauderdale, September 2017; description of art and craft contest and trading of remarks from observer notes; reaction of boys to announcement of winners from observer notes; description of tent vandalisation, reaction of boys and staff from observer notes, interview with OJ Harvey, May 2010, Kelman, H, 'Supplementary Observations', Thursday 6 August 1953, and interview with Prof. Herbert Kelman, July 2012, dialogue up to 'come see' added; dialogue 'Do you see why', 'Laurence is no liar', 'Where were you?' between boys reported in 'Supplementary Observations'; dialogue between OJ and Kelman reported in 'Supplementary Observations'; confrontation between Sherif and Sussman and Harvey's intervention with block of wood from interview with OJ Harvey, May 2010; 'greedy vulture' dialogue added based on Sussman's account of why Sherif was angry with him from letter M Sussman to M Sherif, 30 September 1954, MCWS Papers and interview with OJ Harvey, May 2010; 'Sussman was to go to Panthers' tent and wreck it' from observer notes; Sherif's reaction and announcement that study was over from observer notes, interview with OJ Harvey, May 2010; how men used remaining days and conversation between White, Harvey and Sherif about next study from interview with OJ Harvey, May 2010.

**7: The Robbers Cave**

Driving incident with Sherif at the wheel from interview with OJ Harvey, May 2010; background on Herbert Kelman from Nicholasen, 2017; Sussman wanting to publish study from letter M Sussman to M Sherif, 30 September 1954, MCWS Papers; letter of complaint about Sherif's study from Rockefeller Foundation to M Sherif, Rockefeller Foundation Papers, 1.2, Series 200S, Box 590, Folder 5049; 'something of a disaster' quote from letter L DeVinney to M Sherif, 28 December 1953; Sherif needing more time from M Sherif to L DeVinney, 31

January 1954, and news of a number of papers to be published from letter M Sherif to L DeVinney, 4 May 1954; Sherif's book as manual for researchers from letter M Sherif to L DeVinney, 23 May 1952, Rockefeller Foundation Papers, 1.2, Series 200S, Box 590, Folder 5049; 'well-received book' quote from L DeVinney interoffice memo, 30 September 1954, Rockefeller Foundation Papers, 1.2, Series 200S, Box 590, Folder 5049; Sherif's dismissal of Freudian psychology as a 'failure' from M Sherif, 'A Preliminary Experimental Study of Group Relations', MCWS Papers; Harvey's dissertation, see Harvey, 1953; observing boys in playground from interview with OJ Harvey, May 2010; boys' behaviour on bus from interview with OJ Harvey, May 2010, observer notes, MCWS Papers; background to new territory from Grann, 2017; Robbers Cave built by local prisoners from Perry, Gene, 2012; caretaker finding gold loot in stream from Dyer, 1952; park brochure from Rockefeller Foundation Papers, 1.2, Series 200S, Box 590, Folder 5051; activities of boys on first day from interview with OJ Harvey, May 2010; boys' exploration of cave and surrounds from interview with OJ Harvey, May 2010, dialogue between boys and Hood added; Red inside cave and setting pace for activities from interview with OJ Harvey, May 2010; nightly staff meetings from interview with OJ Harvey, May 2010; camping trip, shooting of snakes from interviews with Bill Snipes, January 2013, Smut Smith, August 2013, OJ Harvey, May 2010, with 'look at that' exchange added; differences between Hollis and Red from observer notes, interview with OJ Harvey, May 2010; Red 'roughing up' and handing out jobs to smaller boys from Sherif et al., 1988, observer notes, and interview with OJ Harvey, May 2010; description of Davey from interview with OJ Harvey, May 2010; Bert Fay's interest in micropaleontology and geology from emails from Bruce Fay, December 2016, and Joyce Stiehler, Oklahoma Geological Survey, July 2013; construction of rope bridge from observer notes, interview with Dwayne Hall, August 2013, 'too fat' dialogue added; killing of copperhead from observer notes, interview with Dwayne Hall, August 2013; Bert Fay and talk of animals from interview with OJ Harvey, May 2010, tent dialogue added based on observer notes of homesickness, and interviews with OJ Harvey, May 2010, and Dwayne Hall, August 2013.

**8: Nation States**

Description of staff meeting, trip to moonshiner from interview with OJ Harvey, May 2010; reaction of both groups to newcomers from observer notes, MCWS Papers, interview with OJ Harvey, May 2010; differing descriptions of Rattlers and Eagles and norms from observer notes, interview with OJ Harvey May 2010, Sherif et al. 1988; Davey's birthday party from interview with OJ Harvey, May 2010; announcement of tournament and reactions of Eagles from audio recordings, MCWS Papers; 'My Carolyn' from letter M Sherif to C Sherif, n.d., MCWS Papers; OJ keeping Sherif busy from interview with OJ Harvey, May 2010; account of ballgame, dialogue between teams, discussion in Eagles' cabin afterwards from observer notes; interviews with Dwayne Hall, August 2013, and OJ Harvey, May 2010; flag-burning incident and dialogue, reaction of Rattlers the next day, Eagles praying before game, and excitement at winning from observer notes, with body language added; Dwayne leading prayer and 'prayers answered' dialogue added based on observer notes, interview with Dwayne Hall, August 2013; Rattlers 'very low' quote from observer notes; hurried meeting about frustration episode from interview with OJ Harvey, May 2010; 'unfair tactics' and 'favorable to a raid' quotes from Sherif et al., 1988; description of night raid and dialogue from observer notes, interview with Bill Snipes, August 2013; reaction and distress of Eagles from observer notes; Eagles dumping mud from interview with OJ Harvey, May 2010; announcement of winners from audio recording, MCWS Papers; Eagles kissing trophy and joy at winning from observer notes; 'sportsmanship giving way' quote from Sherif et al., 1988; Eagles giving three cheers from interviews with OJ Harvey, May 2010, and Dwayne Hall, August 2013; impromptu staff meeting from interview with OJ Harvey, May 2010; 'shouts first' quote from email correspondence with Cindy Lee White, September 2017; men putting 'brakes' on Rattler raid and distress of Eagles after fight from observer notes; 'No (Public) Comment' quote from *Human Events Newsletter* and newspaper reports from Rockefeller Archive Centre, RF 1.2, 200S, Box 590, Folder 5049, Rockefeller Foundation; 'sheer invention' quote from L DeVinney interoffice memo, 28 December 1953, Rockefeller Papers, Box 590, Folder 5050; influence of Cold War from Maekawa, 2009; men reading boys' mail from interview with OJ Harvey, May 2010.

**9: Sweet Harmony**

'Sweet harmony' quote, letter from C Sherif to M Sherif, 1 July 1954, MCWS Papers; conversation between Hood and Rattlers about bet from observer notes; 'after eating for a while', descriptions of 'epic struggle', from Sherif et al., 1988; hostility between groups from observer notes; Ida Bloxham threatens to quit from interview with OJ Harvey, May 2010, observer notes; Ida lays down the law from interview with OJ Harvey, May 2010, observer notes; description of fight with Red and Franklin from observer notes, detail of comic added; Hood breaking up fight, Red running away and hiding, from interview with OJ Harvey, May 2010; dialogue 'we'll bring him back' from observer notes; Hollis and Bill bringing Red back and 'kangaroo court' and expulsion of Red from interview with OJ Harvey, May 2010; description of blocking of water supply, *Treasure Island*, and camping from interview with OJ Harvey, May 2010; trip to Arkansas, announcement about truck, boys drinking soda from observer notes; 'dearest' quote from postcard

M Sherif to C Sherif, n.d, MCWS Papers; photo of boys at state line from MCWS Papers; description of boys in diner at Waldron from observer notes; 'Oklahoma' sung on bus and description of bus trip from Sherif et al., 1988; Hood taking photos of raids from observer notes, interview with OJ Harvey, May 2010, interview with Bill Snipes, August 2013; 'line of action', 'infinite number of events', 'in testing our main hypotheses' quotes from Sherif et al., 1988; 'the competitive situation consists of games' quote from letter M Sherif to L DeVinney, 20 December 1953, Rockefeller Papers, Box 590, Folder 5050.

**10: Empire**

Sherif and camel incident from interview with Sue Sherif, October 2014; description of Charkirge, from Lawrence, 1966; details of Sherif's father and grandparents, Sherif's rivalry with Asch, from interview with Sue Sherif, October 2014; description of Ottoman childhood from Fortna, 2016; wars and tensions between Greeks and Turks, see Zürcher, 2004, Stone, 2010, Milton, 2008, Mansel, 2011; description of town crier from Peterson, 2004; deportation of Armenians from Ödemiş from Batur, 'Young Scientist', 2015; 'these sacred lands' and 'they behaved like worms' quotes and analysis of school readers from Enacar, 2007; Sherif's trip to Smyrna from interview with Sue Sherif, October 2014; descriptions of Smyrna, people, atmosphere, streetscape and avoiding sultan's anger from Milton, 2008, Mansel, 2011; exiled valis and euphoria at end of Abdulhamid's rule from Kechriotis, 2010; Sherif's uncle finding him from interview with Sue Sherif, October 2014.

**11: Burning Memory**

Description of International School from Lawrence, 1966 and Levantine Heritage website; account of tensions between staff at Sherif's school from Ralph Harlow papers; diary excerpt about arrival of British ships after the armistice from Holton, 1918; 'at college mournful groups' quote by Selma Ekrem from Fortna, 2015; background on Poslu Mestan Efe from http://www.erolsasmaz.com/?oku=1568, http://www.haberhurriyeti.com/poslu-mestan-efe-91744.html; accounts of events immediately preceding and including Smyrna fire from Mansel, 2011; brigands attacking headmaster from McLachlan, 1937; Lausanne treaty and forced exchanges from Stone, 2010; collective forgetting of fire from Kolluoğlu-Kırlı, 2005; 'profoundly affected' quote from Sherif, 1967; description of Atatürk's visit to Kastamonou from Seal, 1995.

**12: America and Back**

Sherif's father, description of gravesite, and arrival in America from interview with Sue Sherif, October 2014; hint at temperament from letters between Beryl Parker and Gordon Allport, 1933–1934, Allport Papers, courtesy of the Harvard University Archives (and all subsequent quotes from Allport papers courtesy of Harvard University Archives); background on Cantril from correspondence and interview with Jeff Pooley, January 2017, and 'pathologically ambitious' quoted in Pooley and Socolow, 2013; quotes about Harvard 'blue bloods' and 'read or write' from Granberg and Sarup, 1992; anti-Turk sentiment in America from Grabowski, 2002; Turks visiting Maine from Yalman, 1956; impact of perceived racism on Sherif from letter B Parker to G Allport, 22 January 1934, Allport Papers; 'He is inclined to work too hard' quote from letter G Allport to Prof. M Sekip, 5 January 1931, Allport Papers; interest in range of subjects from Granberg and Sarup, 1992; Lewin and 'fresh breeze' quote from Sherif, 1968; Sherif spending his allowance on books and leaving debts behind at Harvard from letter M Sherif to G Allport, 4 April 1932, Allport Papers; 'sometimes it seems to me' and 'it is a nice dream' quotes from letter M Sherif to G Allport, 14 April 1932, Allport Papers; Sherif's experiences of Germany from letters M Sherif to G Allport, Allport Papers; Menemen incident as watershed from Armstrong, 2016; description of autokinetic experiment and presentation of it to Allport from Routh, 2015; Allport letter describing Sherif as 'unusually original', 'ambitious', 'new framework' from G Allport to Prof. Poffenberger of Columbia University, 20 March 1934; 'leaves temperament behind' from G Allport to B Parker, n.d., Allport Papers; 'Sherif ... was over to dinner' quote by Asa Jennings from Batur, 2015; Sherif's idea of marrying Mary Alice Eaton, her help in writing book from letter M Sherif to G Allport, 19 March 1940, Allport Papers; description of Father Divine and lyrics from H Cantril and M Sherif, 1938; 'last night I attended two meetings' quote from letter M Sherif to Gardner Murphy, 4 September 1934, MCWS Papers; influence of Depression on political views of social scientists from Schrecker, 1986; background on politics at Ankara University from Birkalan, 2001; quote on Sherif's ultra nationalist sympathies from Dinçşahin, 2015, p. 63; account of Sherif's public criticisms of race superiority in Turkey and its effects from Batur, 'Young Scientist', 2015; 'hated and feared' quote from Pratt, 1970; 'inappropriate relationships' quote from Routh, 2015; Sherif protecting Jewish student from letter D Granberg to S Moscovici, 15 April 1986, MCWS Papers; 'failed' quote from letter M Sherif to G Allport, 2 March 1945, Allport Papers.

**13: Oklahoma**

Sherif feeling like Rip van Winkle from letter M Sherif to G Allport, 2 March 1945, Allport Papers, courtesy of the Harvard University Archives; 'beauty and logic' quote from letter C Sherif to M Sherif, n.d., MCWS Papers; Sherif congratulating himself and 'center of my universe' quote from Routh, 2015; 'queer and appalling' quote from letter M Sherif to Carolyn's parents, n.d., from MCWS Papers; 'new sparkle', 'mid Western beauty' quotes from M Sherif letter to unnamed friend, 17 January 1946, MCWS Papers; 'grimly realistic' quote from M Sherif to G Murphy, n.d., MCWS Papers; 'marry an intellectual' from Sherif, C, 1983; 'in order not to strain our relationship' from letter

M Sherif to H Cantril, 16 April 1946, MCWS Papers; background to Cantril, from Glander, 2000; 'Spies, Nazis and SOBs' quote from letter M Sherif to C Pratt, 20 March 1948, MCWS Papers; 'my sweetheart' quote from letter C Sherif to M Sherif, n.d. 1947, MCWS Papers; Sherif's impatience with bureaucracy from interview with OJ Harvey, May 2010; missing appointments in immigration files, MCWS Papers; arrest of Ankara faculty from VanderLippe, 2005; reference to 1949 study as a 'mess' from letter C Sherif to M Sherif, n.d., MCWS Papers; 'head over heels' quote from letter S Flowerman to M Sherif, 19 October 1949, 'shock treatment' quote from letter M Sherif to S Flowerman, 30 October 1949, MCWS Papers; draft paper on 1949 study including descriptions of Bull Dogs and Red Devils from MCWS Papers; influence of Cold War on Sherif from Asliturk and Cherry, 2003; Sherif rejecting fallout shelter money from Granberg and Sarup, 1992; the Sherifs burning books and papers from interview with Prof. Don Granberg, December 2015, and Joan Sherif, September 2017; FBI investigation of SPSSI from Harris, 1980; Sherif's talk at Columbia from Sherif, 1951; FBI investigation of M Sherif from Batur, 'Muzafer Sherif in FBI Files', 2015; FBI investigation of H Cantril from FBI files; Sherif as 'big fish' at OU and Carolyn 'greatest fan' and feedback on presentations from Granberg and Sarup, 1992; description of Sherif as a teacher and Moscovici quote from letter D Granberg to S Moscovici, 15 April 1986, MCWS Papers; relationship between Harvey and Sherif from interview with OJ Harvey, May 2010.

**14: The Museum of Innocence**

'The more I think about it' quote from letter M Sherif to OJ Harvey, 27 December 1975, from OJ Harvey personal papers; 'the crowning one' from M Sherif talk recorded at North Dakota University, 1967, MCWS Papers; Sherif's view of 'the way things work in the real world' and take on experimentation from Granberg and Sarup, 1992, p. 11; 'conflict suffering and agony' from M Sherif, 1967; 'widespread violence' quote from Neyzi, 2008; Hood and 'Pygmalion complex' from interview with Dr Virginia O'Leary, May 2015; incident with Bob Hood and gun from interview with Dr J Trimble; 'not ready for publication' quote from letter, J Lees to M Sherif, 1 February 1955, MCWS Papers; request for reference regarding Sherif from letter L Swearingen to G Allport, 25 January 1960, from Allport Papers; 'eager-beaver' quote from letter G Allport to L Swearingen, 1 February 1960; the Sherifs' generosity to students from Granberg and Sarup, 1992; engaged couples anecdote from interview with Prof. Don Granberg; Sherif's grudges and sensitivity to criticism from Granberg and Sarup, 1992, and carrying critical review from letter D Granberg to S Moscovici, 15 April 1986, MCWS Papers; Sherif's jealousy from Routh, 2015; Sherif's deteriorating health, eating dog food and drinking from Granberg and Sarup, 1992; account of Sherif's car accident from police report, 25 November 1968, MCWS Papers; 'delusions' quote from C Sherif notes, November 1969 (no specific date), MCWS Papers; Carolyn's notes on dream, 1 February 1970, MCWS Papers; Carolyn's career flourishing from Routh, 2015; Carolyn and 'token woman' from C Sherif, 1983; Sherif's embarrassment about his Turkish language from interview with Ann Sherif conducted by Ersin Asliturk, 16 September 2016; folklorist Pertev Boratav and conference cancellation from Öztürkmen, 2005; 'citizen of the world' quote from interview with Sue Sherif; awards and Sherif's complaints about them and less successful versions of textbook from Granberg and Sarup, 1992; Hood teaching from Sherif textbook from interview with Dr V O'Leary, May 2015; CIA funding of Sherif's research and 'did not share consensus' from Marks, 1978; quote from Jay Schulman from Greenfield, 1977; 'academia made room for me' from C Sherif, 1983; Carolyn's efforts to keep him socialising, hospitalisations and suicide attempts from Granberg and Sarup, 1992;

'so much beauty' quote from letter M Sherif to CW Sherif, n.d. 1954, MCWS Papers; talking social psychology from interview with Sue Sherif; reprinting and distribution of book from letter B Hood to M Sherif, 1977; stream of book proposals from MCWS Papers; lithium and its effect on Sherif from interview with Sue Sherif; Sherif's late life trip to Robbers Cave from Granberg and Sarup, 1992; 'OJ old friend' quote from letter M Sherif to OJ Harvey, 20 April 1987, OJ Harvey personal papers; background to Ottoman bandits from Soyudoğan, 2011; picnic with bandits on Mt Bozdağ from Lawrence, 1966; 'first Turk you ever saw' quote from miscellaneous personal correspondence, n.d., MCWS Papers.

# References

## Archives

Alexander McLachlan papers, Queen's University Archives, Kingston, Ontario.

Carl Hovland papers, Manuscripts and Archives, Yale University Library, New Haven, Connecticut.

Faculty Files, University of Utah Archives, Salt Lake City, Utah.

George Lynn Cross papers, Western History Collections, University of Oklahoma Libraries, Norman, Oklahoma.

Gordon Allport papers, Harvard University, Cambridge, Massachusetts.

Hadley Cantril FBI files, National Archives of the FBI, Washington, DC.

Mennonite Historical Library, Goshen College, Goshen, Indiana.

Muzafer and Carolyn Wood Sherif Papers, Archives of the History of American Psychology, The Drs Nicholas and Dorothy Cummings Center for the History of Psychology, The University of Akron (MCWS Papers).

S. Ralph Harlow papers, Amistad Research Center, New Orleans, Louisiana.

William Golding papers, University of Exeter, Exeter, Devon.

Workers Defense League papers, Walter P. Reuther Library Archives of Labor and Urban Affairs, Wayne State University, Detroit, Michigan.

University of Oklahoma Files, Rockefeller Foundation Archives, Rockefeller Archive Centre, Sleepy Hollow, New York.

## Books

Adivar, HE, *Memoirs of Halidé Edib*, 1st edn, J. Murray, London, 1926.

Ahmad, F, *Turkey: the quest for identity*, Oneworld, Oxford, 2003.

Apfelbaum, E, 'Against the Tide: making waves and breaking silences', in Leendert P Mos, (ed.), *History of Psychology in Autobiography*, Springer, New York, 2009.

Baker, WJ, *Playing with God: religion and modern sport*, Harvard University Press, Cambridge, 2007.

Batur, S, 'A Young Scientist in a Changing World: Muzafer Sherif's early years' in A Dost-Gozkan & DS Keith (eds), *Norms, Groups, Conflict, and Social Change*, Transaction Publishers, New Brunswick, New Jersey, 2015.

Batur, S, 'Muzafer Sherif in FBI Files', in A Dost-Gozkan & DS Keith (eds), *Norms, Groups, Conflict, and Social Change*, Transaction Publishers, New Jersey, 2015.

Blackwelder, JK, *Electric City: General Electric in Schenectady*, Texas A&M University Press, College Station, 2014.

Bosmajian, HA, *Burning Books*, McFarland and Company, Inc., North Carolina, 2006.

Carey, J, *William Golding: the man who wrote* Lord of the Flies*: a life*, Faber and Faber, London, 2009.

Cherry, F, *The 'Stubborn Particulars' of Social Psychology: essays on the research process*, Routledge, London and New York, 1995.

Dinçşahin, S, *State and Intellectuals in Turkey: the life and times of Niyazi Berkes, 1908–1988*, Lexington Books, Lanham, 2015.

Dobkin, MH, *Smyrna 1922: the destruction of a city*, Faber, London, 1972.

Doty, RS, *The Character Dimension of Camping*, Association Press, New York, 1960.

Duygu, K, 'Escaping to Girlhood in Late Ottoman Istanbul: Demetra Vaka's and Selma Ekrem's childhood memories', in BC Fortna (ed.), *Childhood in the Late Ottoman Empire and After*, Brill, Leiden and Boston, 2015.

Ekrem, S, *Unveiled: the autobiography of a Turkish girl*, Cultures in Dialogue Series One, Gorgias Press, Piscataway, 2005.

Faroqhi, S, *Subjects of the Sultan: culture and daily life in the Ottoman Empire*, IB Tauris, London, 2000.

Fawaz, L and Bayly, CA (eds), *Modernity and Culture from the Mediterranean to the Indian Ocean, 1890–1920*, Columbia University Press, New York, 2002.

Fortna, BC (ed.), *Childhood in the Late Ottoman Empire and After*, Brill, Leiden and Boston, 2016.

Garnett, LMJ, *Turkish Life in Town and Country*, G Newnes, London, 1904.

Gass, MA, Gillis, HLL and Russell, KC, *Adventure Therapy: theory, research, and practice*, Brunner-Routledge, New York, 2012.

Glander, TR, *Origins of Mass Communications Research During the American Cold War: educational effects and contemporary implications*, L Erlbaum, Mahwah, 2000.

Granberg, D and Sarup, G, *Social Judgment and Intergroup Relations: essays in honor of Muzafer Sherif*, Springer, New York, 1992.

Grann, D, *Killers of the Flower Moon: the Osage murder and the birth of the FBI*, Doubleday, New York, 2017.

Haney, DP, *The Americanization of Social Science Intellectuals and Public Responsibility in the Postwar United States*, Temple University Press, Philadelphia, 2008.

Hanoum, Z, *A Turkish Woman's European Impressions*, Seeley, Service & Co. Ltd., London, 1913.

Hasanli, J, *Stalin and the Turkish Crisis of the Cold War, 1945–1953*, Lexington Books, Lanham, 2011.

Kechriotis, V, 'Celebration and Contestation: the people of Izmir welcome the second constitutional era in 1908', in K Lappas, A Anastasopoulos and E Kolovos (eds), *Μνήμη Πηνελόπης Στάθη: Μελέτες Ιστορίας και Φιλολογίας* Panepistimiakes Ekdeseis Kritis, Irakleio, 2010.

King, C, *Midnight at the Pera Palace: the birth of modern Istanbul*, 1st edn, WW Norton & Company, New York, 2014.

King, GT, *An Autumn Remembered: Bud Wilkinson's legendary '56 Sooners*, University of Oklahoma Press, Norman, 2006.

Korn, JH, *Illusions of Reality: a history of deception in social psychology*, State University of New York Press, Albany, 1997.

Lamprou, A, *Nation-building in Modern Turkey: the 'people's houses', the state and the citizen*, IB Tauris, London, 2015.

Lemov, RM, *World as Laboratory : experiments with mice, mazes, and men*, 1st edn, Hill and Wang, New York, 2005.

Levy, J, *Freudian Slips: all the psychology you need to know*, Michael O'Mara books, London, 2013.

Mango, A, *Atatürk*, John Murray, London, 1999.

Mansel, P, *Levant Splendour and Catastrophe on the Mediterranean*, Yale University Press, New Haven, 2011.

Marks, J, *The Search for the Manchurian Candidate: the CIA and mind control: the secret history of the behavioral sciences*, Norton, New York, 1978.

Milton, G, *Paradise Lost: Smyrna 1922: the destruction of Islam's city of tolerance*, Sceptre, London, 2008.

Nicholson, IM, *Inventing Personality: Gordon Allport and the science of selfhood*, American Psychological Association, Washington, DC., 2003.

Orga, I, *Portrait of a Turkish Family*, Macmillan, New York, 1957.

Pandora, K, *Rebels Within the Ranks: psychologists' critique of scientific authority and democratic realities in New Deal America*, Cambridge University Press, Cambridge and New York, 1997.

Perry, G, *Behind the Shock Machine: the untold story of the notorious Milgram psychology experiments*, Scribe Publications, Melbourne, 2012.

Peterson, MD, *'Starving Armenians': America and the Armenian genocide, 1915–1930 and after*, University of Virginia Press, Charlottesville, 2004.

Pope, N and Pope, H, *Turkey Unveiled: Atatürk and after*, John Murray, London, 1997.

Richards, G, *Putting Psychology in its Place: an introduction from a critical historical perspective*, Routledge, London and New York, 1996.

Rohrer, JH and Sherif, M, *Social Psychology at the Crossroads*, Harper, New York, 1951.

Routh, D, 'The Sherifs' Long Quest to Keep Social Psychology Social', in DS Keith and A Dost-Gozkan (eds), *Norms, Groups, Conflict and Social Change*, Transaction Publishers, New Jersey, 2015.

Schrecker, E, *No Ivory Tower: McCarthyism and the universities*, Oxford University Press, New York, 1986.

Seal, J, *A Fez of the Heart: travels around Turkey in search of a hat*, Picador, London, 1995.

Sherif, C, 'Carolyn Wood Sherif Autobiography', in AN O'Connell and NF Russo (eds), *Models for Achievement: reflections of eminent women in psychology*, Columbia University Press, New York, 1983.

Sherif, M, *In Common Predicament: social psychology of intergroup conflict and cooperation*, International series in the behavioral sciences, Houghton Mifflin, Boston, 1966.

Sherif, M, 'Light from Psychology on Cultural Groups and Their Relations,' in K Bigelow (ed.), *Cultural Groups and Human Relations*, Columbia University Press, New York, 1951.

Sherif, M, *Social Interaction: process and products: selected essays of Muzafer Sherif*, Aldine, Chicago, 1967.

Sherif, M, Harvey, O, Hood, WR, Sherif, CW and White, J, *The Robbers Cave Experiment: intergroup conflict and cooperation*, Wesleyan University Press, Middletown, 1988.

Sherif, M and Murphy, G, *The Psychology of Social Norms*, Harper & Brothers, New York and London, 1936.

Sherif, M and Sherif, CW, *Groups in Harmony and Tension: an integration of studies on intergroup relations*, Octagon Books, New York, 1966.

Sherif, M and Sherif, CW, *Social Psychology*, Harper & Row, New York, 1969.

Simpson, C, *Science of Coercion: communication research and psychological warfare, 1945–1960*, Oxford University Press, New York, 1994.

Somel, SA, *The Modernization of Public Education in the Ottoman Empire, 1839–1908: Islamization, autocracy, and discipline*, Brill, Leiden and Boston, 2001.

Stone, N, *Turkey: a short history*, Thames & Hudson, London, 2010.

Van Slyck, A, *A Manufactured Wilderness: summer camps and the shaping of American youth, 1890–1960*, University of Minnesota Press, Minneapolis, 2006.

VanderLippe, JM, *The Politics of Turkish Democracy: İsmet İnönü and the formation of the multi-party system, 1938–1950*, SUNY Series in the Social and Economic History of the Middle East, State University of New York Press, Albany, 2005.

Yalman, AE, *Turkey in My Time*, University of Oklahoma Press, Norman, 1956.

Yılmaz, H, *Becoming Turkish: nationalist reforms and cultural negotiations in early republican Turkey, 1923–1945*, Syracuse University Press, Syracuse, 2013.

Zürcher, EJ, *Turkey: a modern history*, 3rd edn, IB Tauris, London and New York, 2004.

## Journal and newspaper articles

Aslıtürk, E and Cherry, F, 'Muzafer Sherif: the interconnection of politics and profession', *History and Philosophy of Psychology Bulletin*, vol. 15, no. 2, 2003, pp. 11–6.

Batur, S and Aslıtürk, E, 'On Critical Psychology in Turkey', *Annual Review of Critical Psychology*, 2006, pp. 21–41.

Beck, HP, Levinson, S and Irons, G, 'Finding Little Albert: a journey to John B Watson's infant laboratory', *American Psychologist*, vol. 64, no. 7, 2009, pp. 605–14.

Billig, M, 'Repopulating the Depopulated Pages of Social Psychology', *Theory & Psychology*, vol. 4, no. 3, 1994, pp. 307–35.

Birkalan, H, 'Pertev Naili Boratav, Turkish Politics And The University Events', *Turkish Studies Association Bulletin*, vol. 25, no. 1, 2001, pp. 39–60.

Brannigan, A, 'The Postmodern Experiment: science and ontology in experimental social psychology', *British Journal of Sociology*, vol. 48, no. 4, Dec 1997, pp. 594–610.

Cantril, H and Sherif, M, 'The Kingdom of Father Divine', *Journal of Abnormal and Social Psychology*, vol. 33, 1938, pp. 147–67.

Chamberlin, J, 'Notes on a Scandal', *Monitor on Psychology*, vol. 43, no. 9, 2012, pp. 20.

Cherry, F, 'The Nature Of The Nature of Prejudice', *Journal of the History of the Behavioral Sciences*, vol. 36, no. 4, 2000, pp. 489–98.

Dennis, P, '"Johnny's a gentleman, but Jimmie's a mug": press coverage during the 1930s of Myrtle Mcgraw's study of Johnny and Jimmy Woods', *Journal of the History of the Behavioral Sciences*, vol. 25, 1989, pp. 356–70.

Duman, M Z, 'Trying Not to Forget Forgetting: place of social memory in the construction of the historical and ethnic identity', *Journal of Economic & Social Research*, Vol. 14, issue 2, 2012, pp. 83–99.

Fridlund, AJ, Beck, HP, Goldie, WD, and Irons, G, 'Little Albert: a neurologically impaired child,' *History of Psychology*, vol. 15, issue 4, 2012, pp. 302–27.

Gladstone, AI and Kelman, HC, 'Pacifists vs. Psychologists', *American Psychologist*, vol. 6, no. 4, 1951, pp. 127–8.

Grabowski, J, 'Republican Perceptions, Time and the Gülcemal', *Turkish Yearbook of International Relations*, vol. 30, pp. 31–50.

Greenfield, P, 'CIA's Behavior Caper', *APA Monitor*, December 1977, pp. 110–11.

Greenwood, J, 'What Happened to the "Social" in Social Psychology?', *Journal for the Theory of Social Behaviour*, vol. 34, no. 1, 2004, pp. 19–34.

Harris, B, 'The FBI's Files on APA and SPSSI: description and implications', *American Psychologist*, vol. 35, 1980, pp. 1142–3.

Harris, B, 'Whatever happened to Little Albert?', *American Psychologist*, vol. 34, no. 2, 1979, pp. 151–60.

Harvey, OJ, 'An Experimental Approach to the Study of Status Relations in Informal Groups', *American Sociological Review*, vol.18, no. 4, 1953, pp. 357–367.

James, A B, 'The Life Space Pioneers', *Reclaiming Children and Youth*, vol. 17, no. 2, 2008, pp. 4–10.

Kelman, H, 'Human Use of Human Subjects : the problem of deception in social psychological experiments', *Psychological Bulletin*, vol. 67, no. 1, 1967, pp. 1–11.

Kolluoğlu-Kırlı, B, 'Forgetting the Smyrna Fire', *History Workshop Journal*, vol. 60, Autumn, 2005, pp. 25–44.

Lepore, J, 'Twilight, Growing Old and Even Older', *The New Yorker*, 14 March 2011.

Neyzi, L, 'Remembering Smyrna/Izmir: shared history, shared trauma', *History & Memory*, vol. 20, no. 2, 2008, pp. 106–127.

Nicholson, I, 'The Politics Of Scientific Social Reform, 1936–1960: Goodwin Watson and the society for the psychological study of social issues', *Journal of the History of the Behavioral Sciences*, vol. 33, no. 1, 1997, pp. 39–60.

Nicholson, I, '"Torture at Yale": experimental subjects, laboratory torment and the "rehabilitation" of Milgram's "Obedience to Authority"', *Theory & Psychology*, vol. 21, no. 6, 2011, pp. 737–61.

Öztürkmen, A, 'Folklore on Trial: Pertev Naili Boratav and the denationalization of Turkish Folklore', *Journal of Folklore Research*, vol. 42, no. 2, 2005, pp. 185–216.

Perry, G, 'The View from the Boys', *The Psychologist*, vol. 27, no. 11, 2014, pp. 834–6.

Polansky, N, Lippitt, R and Redl, F, 'An Investigation of Behavioral Contagion in Groups', *Human Relations*, vol. 4, no. 3, 1950, pp. 319–48.

Pooley, J and Socolow, M, 'Checking Up on the Invasion from Mars: Hadley Cantril, Paul F. Lazarsfeld, and the making of a misremembered classic', *International Journal of Communication*, no. 7, 2013, pp. 1920–1948.

Powell, R, Digdon, N, Harris, B and Smithson, C, 'Correcting the record on Watson, Rayner, and Little Albert: Albert Barger as "psychology's lost boy"', *American Psychologist*, vol. 69, no. 6, 2014, pp. 600–11.

Samelson, F, 'From "Race psychology" to "studies of prejudice": some observations on the thematic reversal in social psychology', *Journal of the History of the Behavioral Sciences*, vol. 14, 1978, pp. 265–78.

Sargent, SS and Harris, B, 'Academic Freedom, Civil Liberties, and SPSSI', *Journal of Social Issues*, vol. 42, no. 1, 1986, pp. 43–67.

Schwarzkopf, S, 'Too Many Compromises: survey research and the spectre of communism"', *Journal of Historical Research in Marketing*, vol. 8, no. 1, 2016, pp. 197– 214.

Sherif, M, 'Experiments in Group Conflict', *Scientific American*, vol. 195, 1956, pp. 54–8.

Sherif, M, 'Group Conflict: its course and possible cure', *The Washington Post*, 1 June 1969.

Sherif, M, 'If the Social Scientist is To Be More Than a Mere Technician', *Journal of Social Issues*, vol. 24, 1968, pp. 41–61.

Solovey, M, 'Project Camelot and the 1960s Epistemological Revolution: rethinking the Politics-Patronage-Social Science Nexus', *Social Studies of Science*, vol. 31, no. 2, 2001, pp. 171–206.

Soyudoğan, M, 'Discourse, Identity, and Tribal Banditry: a case study on Ottoman Ayntâb', *International Journal of Turkish Studies*, vol. 17, 2011, pp. 65–93.

Sussman, M, 'A Hero's Journey: the lifeways of a modern day Asclepius', *Marriage & Family Review*, vol. 30, no. 3, 2001, pp. 7–38.

Tomenendal, K, Dogus, F and Mercan, Ö, 'German-Speaking Academic Émigrés in Turkey of the 1940s', *Vertreibung von Wissenschaft*, vol. 21, 2010, pp. 69–99.

Wall, SY, 'Making Modern Childhood, the Natural Way: psychology, mental hygiene, and progressive education at Ontario summer camps, 1920–1955', *Historical Studies in Education/Revue d 'histoire de l'e'ducation*, vol. 20, no. 2, 2008, pp. 73–110.

Winston, AS, 'The Defects of his Race": E. G. Boring and anti-Semitism in American psychology, 1923–1953', *History of Psychology*, vol. 1, no. 1, 1998, pp. 27–51.

## Online articles

Armstrong, W, 'Interview with Şakir Dinçşahin: Niyazi Berkes' life a window onto Turkey's national story', *Hurriyet Daily News*, 20 February 2016, http://www.hurriyetdailynews.com/interview-niyazi-berkes-life-a-window-onto-turkeys-national-story.aspx?pageID=238&nID=95440&NewsCatID=386%3E

Cross, G, 'The University of Oklahoma 1953 – a report', *Sooner Magazine*, vol. 26, no. 2, 1953, pp. 17–21, https://digital.libraries.ou.edu/sooner/articles/p17-21_1953v26n2_OCR.pdf

Dyer, RJ 'State Stream May Be Hiding Bandit's Loot', *The El Reno Daily Tribune*, vol. 61, no. 149, edn 1, 22 Friday 1952, https://gateway.okhistory.org/ark:/67531/metadc920223/m1/1/

Holton, E, 'Segment of the Diary Kept During WWI', *Levantine Heritage*,

August 1918, http://www.levantineheritage.com/diary2.html

Hiltzik, M, 'How a Public Panic Can Cast a Long Shadow: the Nazi summer camp story', *LA Times*, 11 August 2014, http://www.latimes.com/business/hiltzik/la-fi-mh-nazi-summer-camp-story-20140810-column.html%3E

Lawrence, H, 'Memoirs of Helen (Lewis) Lawrence', *Levantine Heritage*, 18 November 1966, http://levantineheritage.com/testi32.htm%3E

Nicholasen, M, 'In Conversation with Herbert C. Kelman: a lifetime in the pursuit of peace', *Centerpiece*, vol. 31, no. 2, 2017, https://wcfia.harvard.edu/publications/centerpiece/spring2017/feature_kelman>.

Perry, Gene, 'A Week With the Boys', *This Land*, 21 March 2012, http://thislandpress.com/2012/03/21/a-week-with-the-boys/

Pratt, C, 'Psychology in America Since 1945', 1970, http://dergiler.ankara.edu.tr/dergiler/26/1610/17334.pdf

Rockefeller Foundation, 'The Rockefeller Foundation Annual Report 1952', The Rockefeller Foundation Library, 13 April 1955, https://assets.rockefellerfoundation.org/app/uploads/20150530122205/Annual-Report-1952.pdf

Weeks, Linton, 'Nazi Summer Camps In 1930s America?', *NPR History Dept*, 28 April 2015, http://www.npr.org/sections/npr-history-dept/2015/04/28/402679062/nazi-summer-camps-in-1930s-america%3E

## Theses

Enacar, Ekin, 'Education, Nationalism and Gender in the Young Turk Era (1908–1918)', Masters thesis, Bilkent University, 2007.

Kendrick, Shelby, '"A Crime Too Terrible for Contemplation": Samuel Ralph Harlow and missionary influence on the history of the responsibility to protect', Bachelor of Arts (Honors) thesis, University of New Orleans, 2014.

Lenser, SD, 'Between the Great Idea and Kemalism: the YMCA at Izmir in the 1920s', Master of Arts in History thesis, Boise State University, 2010.

Maekawa, Reiko, 'The Rockefeller Foundation and the intellectual life of refugee scholars during the Cold War', Kyoto University, 2009.

## Radio programs

Perry, G, *Inside Robbers Cave*, ABC Radio National, 2013.

## Unpublished sources

Kelman, H and Carper, J, 'Factors That Have to be Considered in Evaluating and Discussing Results', 7 August 1953.

MacLachlan, A, *A Potpourri of Sidelights and Shadows from Turkey*, an unpublished memoir, Queens University Archives, 1937.

# Acknowledgements

I could not have undertaken this book without the help of a bevy of people who have generously shared their knowledge, memories, and insights with me. Some of them have their words included in these pages, others have provided crucial background to help me make sense of what happened and why. Firstly there are those people who took part in the experiments themselves, the now-adult 'boys' who gave me permission to share their stories. I would particularly like to thank Doug and June Griset from Schenectady, Bill and Cheri Snipes from Oklahoma City and John and Deb Reich from Phoenix, Arizona for their hospitality and kindness. I'm sorry that OJ Harvey, who lit the spark and encouraged me to pursue this project, never got to see the final result. Thanks to Professor Herbert Kelman, who generously shared his notes and recollections and provided missing pieces.

I could not have embarked on this project without access to Carolyn and Muzafer Sherif's papers and permission from the family to use material from the archives. Special thanks to Sue, Joan, and Ann Sherif and to Professor David Baker, Lizette Royer Barton, and the rest of the staff at The Drs. Nicholas and Dorothy

Cummings Center for the History of Psychology (CCHP) at the University of Akron.

Professor Joseph Trimble went out of his way to lend a hand and tapped his extensive networks numerous times to give me leads, insights, and answers. For research into events in Turkey, I appreciate the work of Dr Sertan Batur, who was unstinting in sharing his ideas and research and patiently answered my numerous questions. I am grateful to Ersin Asliturk for sharing his thoughts on Sherif's theoretical and political orientation, and to Dr M Murat Yurtbilir from the Australian National University, who provided me with historical detail, and Dr Adrian Jones from LaTrobe University, who checked my Turkish research for accuracy.

For background research on Sherif's research, John Reich, Cath and Jerry Felknor, Dr Virginia O'Leary, Cindy White, Professor Michael Lauderdale, Professor Don Granberg, Phil Fox, Hilda Carper and Jean Miller, Professor Dirk Eitzen, and Professor Howard Becker.

Archivist Tuğba Akbörk from Ödemiş Municipality City Museum and Archives gave me terrific insights into Sherif's home town, and archival staff at the University of Exeter were very helpful in giving me access to William Golding's manuscripts and papers.

I could not have begun this book without the careful and inspiring scholarship of the historians of psychology whose work I have drawn on. I was very fortunate that my archival research in Turkey, the United Kingdom, and the United States was supported through the University of Melbourne's Felix Meyer and Lizette Bentwich scholarships. Huge thanks to Drs Kevin Brophy and Rod Buchanan for their encouragement and support, and I'm indebted to the City of Melbourne's Creative Spaces program, who provide me with an office for my research and writing.

Then there were my travel companions. I have Gene Perry (no relation) to thank for a memorable road trip from Tulsa to Robbers Cave, and Faith Wilcox, who helped me find Figsbury Ring, and my cousins, the Shultzes, who provided much-needed relief and companionship on my American visits. Janey Runci has been a wise and constant friend and fellow traveller on the writing journey who has made me laugh and kept me grounded.

I can't really thank my family enough, particularly Dan and Georgia, for putting up with my absences and absent-mindedness with such good humour. I'm fortunate in having Clare Forster from Curtis Brown as my agent and Scribe as my publisher. A huge thanks to Julia Carlomagno, whose faith in this project and insightful suggestions improved it no end.